# THE
# ENGLISH
# FARM  WAGON

MUSEUM OF ENGLISH RURAL LIFE

# THE
# ENGLISH
# FARM WAGON
## ORIGINS AND STRUCTURE

J. GERAINT JENKINS

PUBLISHED BY THE OAKWOOD PRESS FOR THE
UNIVERSITY OF READING
1961

Published by

THE OAKWOOD PRESS

Tandridge Lane, Lingfield, Surrey

for

THE MUSEUM OF ENGLISH RURAL LIFE

UNIVERSITY OF READING

Designed by Margaret Fuller

PRINTED IN ENGLAND
by
A. BROWN & SONS, LIMITED
HULL AND LONDON

# CONTENTS

LIST OF ILLUSTRATIONS . . . . . . . viii

ACKNOWLEDGEMENTS . . . . . . . . x

PREFACE by Andrew Jewell, Keeper, Museum of English Rural Life,
University of Reading . . . . . . . xi

## PART ONE · ORIGINS

### I THE WAGON

The evolution of the four-wheeled wagon and its relation to other
wheeled transport

### II CONSTRUCTION

The evolution of wagon building techniques

1 WHEELS . . . . . . . . . . 22
2 UNDERCARRIAGE . . . . . . . . 33
3 DRAUGHT POLE, SHAFTS AND TRACES . . . . 34
4 WAGON BODIES . . . . . . . . . 37

### III DISTRIBUTION

The geographical areas where the wagon is used and the reason
for its adoption

1 CONTINENTAL DISTRIBUTION . . . . . 43
2 BRITISH DISTRIBUTION . . . . . . . 47
3 REASONS FOR WAGON USE . . . . . . 50
    (1) Topographical Factors . . . . . 50
    (2) Economic Factors . . . . . . 52
    (3) Size of Holdings . . . . . . 53
    (4) Field Patterns . . . . . . . 54
    (5) Cultural Factors . . . . . . 55

# PART TWO · STRUCTURE

## IV BUILDING
### The craft of the English wheelwright

1 WHEELS . . . . . . . . . . 61
    (1) Nave . . . . . . . . 61
    (2) Spokes . . . . . . . . 66
    (3) Felloes . . . . . . . . 68
    (4) Tyre . . . . . . . . 71
    (5) Box . . . . . . . . 80
2 UNDERCARRIAGE . . . . . . . . 81
    (1) Axles . . . . . . . . 81
    (2) Fore and Rear Carriages . . . . . 83
3 SHAFTS . . . . . . . . . . 87
4 BODY . . . . . . . . . . . 89
    (1) Framework . . . . . . . 89
    (2) Sides . . . . . . . . 91
    (3) Side Supports . . . . . . . 94
    (4) Sideboards . . . . . . . 96
    (5) End Boards . . . . . . . 98
5 METHODS OF BRAKING . . . . . . . 99
    (1) Roller Scotches . . . . . . 99
    (2) Dog Sticks . . . . . . . 99
    (3) Drag Shoes . . . . . . . 100
6 LOCK . . . . . . . . . . . 101
7 LADDERS . . . . . . . . . . 103
8 PAINTING AND DECORATIONS . . . . . 104
    (1) Frontboard Designs . . . . . . 106
    (2) Tailboard Designs . . . . . . 109
    (3) Side Decorations . . . . . . 110

## V BOX WAGONS

1 EASTERN COUNTIES WAGONS . . . . . 114
    (1) Lincolnshire Wagon . . . . . . 114
    (2) East Anglian Wagon . . . . . 119
2 EAST MIDLANDS WAGONS . . . . . . 125
    (1) Hertfordshire Wagon . . . . . 125
    (2) Rutland Wagon . . . . . . 130
3 WEST MIDLANDS WAGONS . . . . . . 134
(a) SOUTHERN TYPE . . . . . . . . 135
    (1) Hereford Panel-Sided Wagon . . . . 135
    (2) Hereford Plank-Sided Wagon . . . . 139
    (3) Worcestershire Wagon . . . . . 140
    (4) Monmouthshire Wagon . . . . . 142
    (5) Radnorshire Wagon . . . . . . 144

(b) CENTRAL TYPE . . . . . . . . 146
      (1) Shropshire Wagon . . . . . 146
      (2) Montgomeryshire Wagon . . . . 150
(c) NORTHERN TYPE . . . . . . . . 152
      (1) Staffordshire Wagon . . . . . 152
      (2) Denbighshire Wagon . . . . . 155
4 SOUTH-EASTERN WAGONS . . . . . . 157
      (1) Sussex Wagon . . . . . . 158
      (2) Kent Wagon . . . . . . . 163
5 CENTRAL SOUTHERN ENGLAND WAGONS . . . 164
      (1) Surrey Wagon . . . . . . 165
      (2) Dorset Wagon . . . . . . . 169
6 YORKSHIRE WAGON . . . . . . . 173


VI BOW WAGONS

1 SOUTH MIDLANDS SPINDLE-SIDED WAGON . . 182
2 WESSEX AND THE LOWER SEVERN BASIN PANEL-SIDED
   WAGONS . . . . . . . . . 188
      (1) Wiltshire Wagon . . . . . . 189
      (2) West Berkshire Wagon . . . . . 194
      (3) North-West Hampshire Wagon . . . . 195
      (4) Dorset Bow Wagon . . . . . . 197
      (5) North Somerset and Vale of Berkeley Wagon . 198
      (6) Glamorgan Wagon . . . . . . 200
3 SOUTH-WESTERN WAGONS . . . . . . 204
      (1) Somerset Wagon . . . . . . 205
      (2) Devon Wagon . . . . . . . 209
      (3) Cornish Wagon . . . . . . 211


APPENDICES

I Measurements of Examples of Each Wagon Type . . . 217
II Features of Construction of Each Wagon Type . . . . 230
III Wagon Types in Each County . . . . . . 232
IV Catalogue of Wagons in Museums . . . . . . 235


INDEX . . . . . . . . . . . 243

# ILLUSTRATIONS

1    Triangular ox cart from Sind (after Haudricourt)     .     .     .     4
2    Representation of a wagon from a cave drawing at Langon, Sweden
      (after Berg) .    .    .    .    .    .    .    .    5
3    Dutch wagon, *Siege of s'Hertogenbosch* by Van Hillegaert, late
      sixteenth century    .    .    .    .    .    .    9
4    (a) Carrier's wagon from an engraving by W. H. Pyne dated 1802    10
      (b) Carrier's wagon of 1780 from Streetly End, Cambridgeshire .    11
5    Cornish wain    .    .    .    .    .    .    .    .    14
6    Barge wagon .    .    .    .    .    .    .    .    .    16
7    Boat wagon    .    .    .    .    .    .    .    .    .    17
8    Wheel from the Roman frontier post at Newstead    .    .    .    26
9    The dished wheel (after Sturt) .    .    .    .    .    .    28
10    Map showing the distribution of carts and wagons in Europe    .    44
11    Modern European wagons:
          (a) U.S.S.R.    .    .    .    .    .    .    .    46
          (b) France .    .    .    .    .    .    .    .    47
12    Map showing the distribution and regional types of English farm
      wagon    .    .    .    .    .    .    .    .    49
13    The wheelwright, an engraving by Stanley Anderson, CBE, RA    .    62
14    Wheel naves: barrel-shaped and modern cylindrical type    .    .    65
15    Saw-pit at Pontrilas, Hereford, 1920 .    .    .    .    .    69
16    Tyring a wheel at Ardington, Berkshire, 1959    .    .    .    72
17    (a) and (b) The wheelwright's tools .    .    .    .    .    76
      (c) A wheelwright's lathe    .    .    .    .    .    .    78
18    Axles: wooden axle with wooden arms and wooden axle with
      iron arms    .    .    .    .    .    .    .    .    82
19    Front elevation of a Bow Wagon    .    .    .    .    .    84
20    Variations in forecarriage design    .    .    .    .    .    86
21    Arrangement of undercarriage in an East Anglian Wagon    .    .    88
22    Body framework of a waisted wagon .    .    .    .    .    92
23    Methods of side construction: (a) Spindle-sided; (b) Spindle-sided
      with midrail; (c) Panel-sided; (d) Panel-sided with midrail;
      (e) Plank-sided    .    .    .    .    .    .    .    95
24    Designs of side supports .    .    .    .    .    .    .    96
25    Decorations on frontboards    .    .    .    .    .    .    107
26    Lincolnshire Wagon from Leadenham    .    .    .    .    115
27    East Anglian Wagon from Sible Headingham, Essex    .    .    120
28    Hermaphrodite from Gillingham, Norfolk .    .    .    .    122
29    Hertfordshire Wagon from Harpenden    .    .    .    .    126
30    Rutland Wagon from Teigh    .    .    .    .    .    .    133
31    Hereford Panel-Sided Wagon .    .    .    .    .    .    136
32    Hereford Plank-Sided Wagon, a drawing by Thomas Hennel    .    139
33    Worcestershire Wagon from Powick .    .    .    .    .    141
34    West Midlands Trolley from Presteigne    .    .    .    .    142

35 Monmouthshire Wagon from Tirley, Gloucestershire . . . 144
36 Radnorshire Wagon from Newchurch . . . . . 145
37 Shropshire Wagon from Plowden . . . . . . 150
38 Montgomeryshire Wagon from Llanwnnog . . . . 151
39 Staffordshire Wagon from Lichfield . . . . . . 153
40 Denbighshire Wagon from Rhuthun . . . . . . 156
41 Sussex Wagon from Horsham . . . . . . . 161
42 Kent Wagon . . . . . . . . . . 163
43 Surrey Wagon . . . . . . . . . 166
44 Dorset Wagon from Sherborne . . . . . . 170
45 Yorkshire Wagon from Carnaby . . . . . . 175
46 South Midlands Spindle-Sided Wagon from Hailey, Oxfordshire . 184
47 Wiltshire Wagon from Pewsham . . . . . . 190
48 West Berkshire Wagon from Boxford . . . . . 194
49 North West Hampshire Wagon . . . . . . 196
50 Dorset Bow Wagon from Plaitford . . . . . . 199
51 North Somerset and Vale of Berkeley Wagon from Dymock . . 200
52 Glamorgan Wagon from Llanishen . . . . . . 202
53 A late example of a Glamorgan Wagon from Pyle . . . 203
54 Somerset Wagon from Durston . . . . . . 207
55 Devon Wagon from Upton . . . . . . . 210
56 Cornish fully-locking Wagon . . . . . . . 212
57 Diagram of the East Anglian Box Wagon with named parts . 214
58 Diagram of the South Midlands Bow Wagon with named parts . 215

The illustrations are reproduced by permission of:

Mr. Stanley Anderson, C.B.E., R.A. (13); Cambridge University Press (32); Castle Museum, York (45); Miss F. Foster (28); Mrs. L. M. Jones (23); The Rev. R. H. Lane (29); Ministry of Agriculture, U.S.S.R. (11a); Museum of Agriculture, Wye College (42); National Museum of Antiquities of Scotland (8); National Museum of Wales, Welsh Folk Museum (34), (36), (38), (40), (52), (53); Het Nederlands Openluchtmuseum, Arnhem (3); Rothamsted Experimental Station (29), (31), (43); Taskers of Andover (6); Miss M. Wight (15); Mr. David Wray (56).

The remaining illustrations are from the records of the Museum of English Rural Life, University of Reading.

# ACKNOWLEDGEMENTS

I wish to express my gratitude to all those who have contributed in various ways to the successful realization of this project. This attempt to make a thorough survey of the English farm wagon would have been impossible without the co-operation and assistance of the very many people who kindly completed questionnaires, supplied information and gave me free access to their fields and barns.

I am especially grateful to the Reverend R. H. Lane, Mr. C. F. Tebbutt and Mr. N. A. Hudleston for allowing me to make use of their own records and to the Curators of the many European and British museums, the County Secretaries of the National Farmers' Union who supplied me with information, The Pilgrim Trust and the University of Reading Research Board. I should like to thank the Keeper and Curators of the Museum of English Rural Life, and Mr. John Higgs and Mr. Antony Easthope for valuable assistance and advice at all stages. I am also indebted to Mr. John Anstee for preparing the line drawings.

Grateful acknowledgement must also be made for the generous donations towards the publication of this book which were given by the Dunlop Rubber Company, Esso Petroleum Company, Fisons, Shell-Mex and B.P., and Taskers of Andover.

J. GERAINT JENKINS

# PREFACE

The Museum of English Rural Life was established to provide a centre of information concerned with the material of the countryside. For this purpose it is not enough merely to collect such things as ploughs, smocks and craftsmen's tools. The countryman's equipment may sometimes have a sentimental value, and it is often curiously contrived, but it seldom has an intrinsic value of its own. To the student who is seeking to reconstruct the daily life of the countryside in the past, these material objects are useful only in proportion to the information that can be gleaned about their background history. This information is vanishing even more quickly than the objects themselves and there is little enough time left in which to gather for future generations that which now remains. Such was the plight of the English farm wagon.

For about two hundred years a village-made four-wheeled vehicle, called a wagon to distinguish it from the cart with only two wheels, was a common sight in many parts of the English countryside. It was used for carrying the hay and corn harvest from the fields and for transporting grain and other produce to market. As long ago as the 1870s Richard Jefferies was writing of the 'true old fashioned wagon' as if it were already disappearing from his 'Southern County', and indeed it was. At that time, factory-made boat and barge wagons had appeared, but the village wheelwrights built their wagons to last more than one farmer's lifetime and it is some tribute to their craftsmanship that the author of this book was able to find examples of traditional wagons in every district of the country in which they had been used. It was not always easy. A week of searching in Staffordshire only brought to light one vehicle in a good state of preservation and this wagon is now in the Museum of English Rural Life. The oldest wagon of all, built in 1780, was found under a roof that had collapsed twenty-five years ago.

This book is based on a survey of nearly 600 wagons which were traced through enquiries in the Press and with the assistance of organizations such as the National Farmers' Union, the Rural

Industries Bureau, regional museums and local history councils. A questionnaire for recording the details of wagon construction was sent to farmers and other local co-operators and the author himself measured and recorded wagons in all parts of the country. From the information provided by the survey, it has been possible for the first time to describe the characteristics of some twenty-eight distinct regional types of English farm wagon and to define the areas in which they were built.

For the story of the evolution of the wagon, outlined in the first part of the book, the author has drawn for comparative material upon the knowledge of workers in most of the countries of Europe in order to trace the early history of the wagon back to prehistoric times. By the end of the eighteenth century, the two main families of English farm wagon, the box and the bow, were well established. The relationship of these to two different cultural influences is one of the more interesting aspects of the author's study.

The English farm wagon is only one small part of the large field which the Museum of English Rural Life has set out to study. Research of this kind cannot be completed in a library or in a museum. The investigator must spend long periods on field work often with the knowledge that the material evidence he is seeking may be destroyed tomorrow and that information is but precariously held in the memories of the oldest members of the community.

The extent of the Museum's success in recording depends so much on the financial support which it receives. Although acknowledgement to all those who have assisted the author is made elsewhere, I particularly wish to thank those organizations which have generously subscribed to the publication of this book, for without their help it could not have been made available to the general public.

UNIVERSITY OF READING                                         ANDREW JEWELL
JUNE 1960

# PART ONE
# ORIGINS

# THE WAGON

## I

The earliest wheeled vehicle was almost certainly a two-wheeled ox cart, of a type in general use among the urban communities of the Middle and Near East as early as 3000 B.C. This vehicle was quite simply constructed with solid wheels and a triangular frame of two poles joined together at the front. The apex was extended to provide a central draught pole to which the ox yoke was lashed. Carts of almost identical design are to be found to this day in some of the remote regions of Spain, Portugal, Sardinia and South-East Asia (Fig. 1).

The wagon was not entirely unknown to the neolithic peoples of the Middle and Near East. Models of both carts and wagons, dating from around 3000 B.C., have been found at Tepe Gawra in Mesopotamia, and both types of vehicle are represented in Syria and the Anatolian Plateau.[1] A clay model of a wagon from Palaikastro points to the existence of four-wheeled vehicles in Crete as early as 1800 B.C.[2]

The prehistoric wagon was simply a rectangular box on four solid wheels, and all the evidence suggests that its use was confined to ceremonial and cult purposes and was very seldom employed for the ordinary transport of goods and people. To this day wagons in Mediterranean Europe are extremely rare, and the few that do exist are used on ceremonial occasions such as the processions on saints' feast days. The ordinary transport around farm and village continues to be performed by the ubiquitous ox cart.

Although four-wheeled vehicles were known in South Europe and South-West Asia in prehistoric times, in construction they are far removed from the wagons of Central and West Europe in later times. One must therefore look elsewhere for the origins of the English farm wagon.

According to the noted Polish anthropologist, Jan Czekanowski, the earliest form of wagon in temperate Europe came about as the result of joining together two triangular-framed ox carts.[3] The scanty evidence suggests that this may have taken place in the Lausitz district of Central Germany during the Late Bronze Age. Undoubtedly the ox cart was well known to the inhabitants of Lausitz at this time, for they had numerous contacts of trade with the urban civilisations of Southern Eurasia.[4] Bronze Age Lausitz was the great cultural melting pot of Central Europe; it was a virile society of farmers, metallurgists and traders who travelled widely and contributed much to the way of life of other European communities from the Balkans to the Atlantic Coast. Undoubtedly it was the Lusatian people that introduced the four-wheeled wagon to all parts of the North European Plain. By the Late Bronze Age the wagon was known in Scandinavia, the evidence for its existence being seen in a rock carving at Langon in Sweden[5] (Fig. 2). A sherd depicting a funeral carriage shows the existence of the wagon at Odenburg in Hungary in Late Halstatt times, while the famous cult wagon excavated from the Dejbjerg Bog in Jutland,[6] points to the existence of simply constructed, but elaborately decorated, wagons in Denmark in the Early Iron Age.

All these early wagons were built up of two identical sections,[7] each section representing a triangular-framed cart. The end of the rear section was joined to the fore-section by means of a coupling pin that passed through the apex of the triangle formed by the converging framework of the rear part, and through the floor of the fore-part. A constructional feature that characterizes all European wagons is the coupling pole that joins the forecarriage to the rear carriage, and this feature, occurring on all wagons from the Bronze Age to the nineteenth century, may well represent the central draught pole of the rear ox cart in a composite vehicle.

In construction the earliest form of wagon had one great dis-

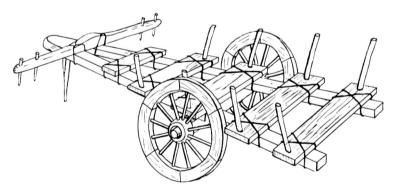

1   Triangular ox cart from Sind (after Haudricourt)

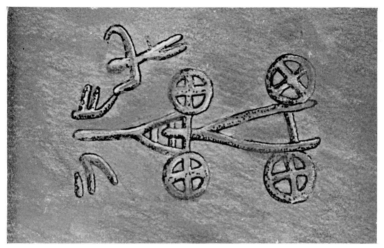

2   A wagon from a cave drawing at Langon, Sweden (after Berg)

advantage. Since the vehicle was built of two identical sections pinned to one another by the frame, the forecarriage was almost immovable in relation to the rear carriage. Like some cheaper varieties of children's toy vehicles of the present day, the whole vehicle had to be dragged around to turn a corner.[8] In some of the more remote regions of Europe, similar wagons equipped with central draught poles and rigid forecarriages may still be seen in constant use. Czekanowski,[9] for example, notes that in his childhood in the Grojec district of Poland such vehicles were widely used, while in Bohemia,[10] in Scandinavia[11] and other places these primitive vehicles, which date back in type to the Late Bronze Age, may still be seen.

Almost until the end of the Roman Empire, the only vehicles known in Europe were those equipped with central draught poles and generally drawn by a pair of oxen in yokes. Although the horse, a native of Central Asia, was known in Bronze Age Europe, it was rarely used for drawing any vehicles but those of ceremony and war. The sherd of pottery from Odenburg, for example, shows a pair of horses harnessed to the draught pole of a funeral carriage, while the Dejbjerg wagon is equipped with a central pole. There is no evidence that any other form of harnessing was known in prehistoric times and, for as long as a rigid extension of the triangular frame of the wagon served as a draught pole, any change in the design of the forecarriage was impossible. The earliest representation of

side shafts is found on the monument at Igiel near Trèves, which
dates from the decline of the Roman Empire, and from that time
the substitution of side shafts for the central draught pole spread
very slowly to many parts of Europe.

The use of side shafts in place of a draught pole came about as
the result of a radical change in the harness of draught animals.
While an ox develops its draught power by pushing against a yoke
over its shoulders, a horse pushes against a collar and a set of hames
around its neck. The method of harnessing horses to a draught pole
is highly unsatisfactory and, at an early date, man attempted to
improve draught by adopting a new method of harnessing. This
radical change in harness was the introduction of the horse collar
into Europe from Central Asia by the Hun invaders.[12]

With the appearance of a new method of harnessing, the construction of the wagon consequently underwent considerable
changes. The central pole which had hitherto connected the vehicle
to the yoke of draught animals became unsatisfactory, now that
horses equipped with collars were widely used. The best method of
attaching a vehicle to the draught animals was by means of side
shafts to which the hames and collar of the horse were connected
by hooks and chains, and it is highly probable that the use of side
shafts in Europe spread at the same time as the collar.[13]

Unlike the central draught pole, which was merely a continuation of the apex of the triangle formed by the frame of the wagon,
side shafts were constructed independently, not as an integral part
of the frame but pinned or bolted to the forecarriage. With the
disappearance of the rigid draught pole and its replacement by
removable shafts, a locking forecarriage appeared, and this method
of construction, which occurs on all modern European wagons,
gradually spread after the fourth century A.D. to all parts of Europe.
Very little is known of the construction of wheeled vehicles between
the fall of the Roman Empire and the close of the Middle Ages.
After the adoption of side shafts, it is unlikely that medieval vehicles
showed any technical advance on those of the Roman world.
Medieval records speak of *carreta*, presumably two-wheeled carts,
and *carreta longa*, the long carts of Piers Plowman. The latter may
have been four-wheeled, but were more likely two-wheeled carts
with long bodies similar to the harvest *wain* of Cornwall and the
*gambo* of Wales.

The Luttrell Psalter of the fourteenth century shows an

elaborately decorated covered wagon which was probably used for carrying people rather than goods. As such it may be regarded as the direct successor of the Roman *carruca*, which was also a covered, long distance sleeping carriage, rather than a predecessor of the carrier's and farmer's wagons of post-medieval Britain. The fact that both the rear and fore-wheels of the vehicle are of equal size, the elaborate carving of the body, the heavy tilt, all suggest little relation to the later wagons.

In addition to travelling carriages, some form of baggage wagon was well known in medieval Britain. A long series of writs dated 1333[14] ordered 'various abbots and priors to send carts and wagons with the necessary horses to carry to the north, tents and other things necessary for the King's expedition against the Scots'.

It is certain therefore that some form of four-wheeled vehicle, whether it be a travelling coach or baggage wagon, was well known in Britain in medieval times, but there is no evidence to suggest that any form of farm wagon was in existence. A true picture of the medieval highway would show men riding on horseback, pack horses with panniers or bundles on their backs, numerous slow moving carts and a very occasional four-wheeled passenger or baggage vehicle. The rarity of four-wheeled vehicles at this time has led some writers to believe that they were not introduced into the country until at least the beginning of the sixteenth century,[15] but the baggage wagon was undoubtedly well known in the more level parts of the country in medieval times. A farm wagon, however, must have been a rarity.

Fitzherbert in his *Book of Husbandry* of 1623 does not mention a wagon as being a part of a farmer's requirements,[16] although he goes into great detail regarding the equipment of the sixteenth-century farmer. He does say, however, that a cart, presumably a box cart or tumbril, and a wain, perhaps similar to the two-wheeled, long-bodied Cornish wain and Welsh *gambo* were essential.[17] The word 'wagon' had already been incorporated in the English language[18] by 1557 when Thomas Tusser wrote *Hundred Points of Good Husbandry*. But even in East Anglia, the region of which Tusser wrote, a region where wagons made an early appearance, four-wheeled vehicles must have been very scarce and carts were in general use for harvesting.[19] Despite the circumspection with which Tusser's doggerel verse must be viewed as historical evidence, it is quite clear that he was not familiar with wagons, at least as part of the equipment of the sixteenth-century farmer in East Anglia.

The sixteenth century has been described as the great turning point in British agriculture.[20] The medieval organisation of society with its trade gilds in the towns and the manorial system of farming in the country slowly changed, and a new tide of vigour, which seemed to awake fresh effort in every aspect of life, swept the country. The object of the elaborately organised medieval village was to produce food and other commodities for a self-sufficing community, but with the great rise of population, especially in the towns, the incentive for increased food production was provided. The aim of the Tudor farmer was not only to supply his own needs, but to raise crops and breed animals which he could sell for a profit, so making the marketing of farm produce increasingly important. The release of initiative, brought about by the replacement of a self-sufficing economy by one of individual farming for profit, led to experiments in the crafts dependent on agriculture. Larger and better vehicles were required not only to transport the harvest from field to farmyard but also to carry the produce of the countryside to the towns. From the mid-sixteenth century the quality of wheel- and wain-wrighting improved greatly and the number of vehicles on English roads increased steadily as the years passed by.

Over much of England carts of greatly improved construction continued to be used for farm transport, but in some areas, notably in East Anglia, a few wagons were introduced for harvesting hay and corn. The greatest change, however, in the means of transport in the sixteenth century was the introduction of heavy wagons for the long distance haulage of goods. These wagons were a new departure from the baggage and passenger wagons of the medieval world, and they were almost certainly introduced into Britain from the Low Countries. In the sixteenth century Britain had numerous trading contacts with the Netherlands, and at the same time a large number of aliens from Holland, Belgium and North France were resident in Britain.[21]

It is to be expected that the construction of the wagons of the Low Countries in the sixteenth century should display many of the characteristics of those in England at that time. A typical Dutch wagon is shown in the painting *Siege of s'Hertogenbosch*, by Van Hillegaert towards the end of the sixteenth century[22] (Fig. 3). This

3 Dutch wagon, *Siege of s'Hertogenbosch* by Van Hillegaert, late sixteenth
century

wagon has two pairs of narrow tyred, greatly dished wheels, the
rear wheels being considerably larger than the fore-wheels. The
naves are elongated and each wheel is equipped with eight spokes.
The body consists of a large number of wooden spindles running
from a straight frame to a top-rail that rises steeply to a deep front
board and less steeply to a tailboard. The wagon is attached to the
two draught horses by a series of chain traces. A large number of
sixteenth and seventeenth-century prints and drawings show
exactly the same type of vehicle in Britain. A well known mid-
seventeenth century print, *A Cart loaded with stone in difficulties on a
rough road*, by W. Faithorne,[23] shows a wagon similar in many
respects to a Dutch vehicle. Its narrow, tyred, dished wheels, the
numerous body spindles, the curved profile and chain traces shows
its very close relationship to the wagons of the Low Countries. It
seems reasonable to suppose, therefore, that the English carrier's
wagon was an adaptation of the continental vehicle, rather than a
natural development from the native, medieval vehicles.

From the mid-sixteenth century the spindle-sided carrier's wagon
became a very common sight on English roads. At first the largest
of these wagons could only carry four tons but, as the years passed
by, wagons became larger and larger so that by the close of the

4 (a) Carrier's wagon, an engraving by W. H. Pyne dated 1802

seventeenth century a carrier's wagon could transport anything up to eight tons in weight. Drawn by a team of up to twelve horses each wagon ran on a fixed route on specified days. A weekly service between a number of Sussex towns and London, for example, was instituted in the late seventeenth century, and each vehicle carried huge loads of wheat, oats and fodder to the metropolis. The ships' timbers from London shipbreaking yards, which they loaded for the return journey, were used by Sussex landowners to build cottages and barns on their estates.[24]

The old road wagons were well suited for this long distance haulage but their great weight made them unsuitable for transport around the farm. A few farm wagons, probably large and unwieldy, were to be found on some of the larger estates from the early seventeenth century,[25] but over the larger part of the country the ubiquitous cart was used for all farm transport. It is certain that the distinct regional varieties of wagon did not evolve until at least the end of the eighteenth century.

During the mid-eighteenth century the whole aspect of British agriculture was changing very quickly. Between 1727 and 1760 more than two hundred private Enclosure Acts were passed and enclosed farming became the rule rather than the exception. With this great change in the agricultural economy, better tools and implements were adopted in great quantities and there was a demand for something much larger than carts for harvesting on English farms. In order to satisfy this demand, village craftsmen began to build large wagons, adapting the design of the old carriers' wagons to the needs of their own locality. Wagons of three or four tons capacity were built in great numbers and these loaded with hay or corn sheaves could be drawn by two horses from field to rick yard.

Although the general principles on which wagons were built were uniform throughout the country, there was a great deal of variation in detailed design. In some districts, such as the East Midlands, the wagon builders stuck very closely to the form of the older road wagons, while in other regions this design was considerably changed to suit the varying demands of soil and topography and so the distinct regional types of farm wagon came into existence. For example, the wagons of East Anglia were heavy and box-like as befitted a level area of large fields. Those of the Cotswolds were light and excess weight was cut down wherever possible in order to make them suitable to the needs of an undulating region with numerous steep slopes. In Sussex broad-wheeled wagons were used in the clay land of the Weald, while narrow-tyred wheels were general on the lighter soils of the Downs.

Even so the adaptation of wagon design to topography and soil only partly explains why distinct regional types came into existence. There are examples of wagons which are not at all well adapted to the district where they are used. In South Shropshire, for example, the wagons are very heavy and broad-wheeled and seem

4   (b)  Carrier's wagon of 1780 from Streetly End, Cambridgeshire

ill-suited to the hilly nature of the countryside, yet these heavy wagons occur with little variation in design and construction throughout the whole of the area. In size and appearance the Shropshire farm wagon resembles the older road wagon, and it seems that, when in the eighteenth century it became the custom to use four-wheeled vehicles for harvesting, no drastic adaptation to local conditions of the existing road wagons took place in the county as in other regions. The traditional pattern continued to be built in Shropshire for as long as village craftsmen constructed farm vehicles.

In some regions the wagons were closely related in design and construction to their two-wheeled predecessors. The East Anglian tumbril, for instance, is a deep-bodied, blue-coloured vehicle with sides rising steeply to a lofty frontboard. Undoubtedly this two-wheeled vehicle, still found in many parts of East Anglia is the direct descendant of the medieval box cart. The East Anglian wagon (Fig. 27), like the tumbril, is deep-bodied, it is painted blue, its panelled sides rise steeply to a lofty frontboard, while the wooden side supports and narrow sideboards are similar in shape to those of the tumbril. In appearance and construction the East Anglian wagon may be regarded as a four-wheeled version of the earlier tumbril, and from the mid-eighteenth to the early twentieth centuries it was built throughout East Anglia with little variation in design.

English farm wagons of the late eighteenth and nineteenth centuries may be divided into two distinct types, the 'box wagon' (Fig. 57) and the 'bow wagon' (Fig. 58). The box wagon has a deep rectangular body and may be regarded as a direct successor of the road wagon of an earlier era, which itself was an English version of the continental vehicle. Some box wagons, like those found in Lincoln (Fig. 26) and the adjacent counties closely resemble those of the Low Countries in general appearance. The detail of the curved top-rails, the large greatly dished wheels and numerous body spindles are, however, derived from the carrier's wagon. Outside Lincolnshire various improved versions of the box wagon are found throughout the south-eastern quadrant of the British Isles from Sussex to Shropshire and from East Anglia to Herefordshire.

Unlike continental wagons, most of the English farm vehicles are equipped with sideboards to take the overhanging load, and this feature, together with the waisting of the side-frames for greater lock, may be regarded as the contribution of English craftsmen to wagon design. The box wagon, therefore, must be regarded as an improved version of the continental vehicle, and not merely a direct copy of some pre-existing pattern.

While the box wagon does show many points of similarity to the wagons of the continent in appearance and construction, the bow wagon is very far removed from them. The main characteristics of the bow wagon are the very wide sideboards that curve in an arch over the large, greatly dished rear wheels, the shallow body and the elegance and light appearance of the whole vehicle.

A great fault of many box wagons is the fact that the wheels are large and the body is far too high for easy loading. The only method of overcoming this would be to lessen the diameter of the wheels to lower the floor of the vehicle. But small wheels are a disadvantage as a vehicle equipped with them has great difficulty in running along rutted tracks and over heavy wet soil. Nineteenth-century craftsmen found it a great problem to strike the correct balance between wheel size and height of wagon body. In the West of England, it was overcome by providing wagons with very wide sideboards that curved over large rear wheels. Although the body of the bow wagon was made shallow, its carrying capacity compared very favourably with most medium sized box wagons, because the overhanging sideboards themselves carried a considerable load.

The geographical distribution of the bow wagon (Fig. 12) is confined to the south-western quadrant of the British Isles, over an area that extends from the Chilterns to the Cornish Peninsula. This limited distribution provides a key to the evolution of the bow wagon for it undoubtedly developed from a two-wheeled harvest cart or wain, used in the West Country for countless centuries, and still found on a number of Cornish farms (Fig. 5). The wain may be regarded as the British representative of the Mediterranean ox cart and in early nineteenth-century Cornwall it was not an uncommon practice to yoke a pair of oxen to it.[26] The body of the wain consists of a long rectangular platform about 13 ft. long and 6 ft. wide. The wheels are about 4 ft. in diameter while a railed arch prevents the overspilling load from clogging them. According to Worgan a two-wheeled wain could carry up to 250 sheaves.[27]

At the beginning of the nineteenth century a few wagons were

also found in Cornwall.[28] In shape and construction they were the
same as the two-wheeled wains. Similarly, in Wales, two-wheeled
carts which are very closely related to the bow wagon are also
found. An eighteenth-century wain preserved at the Welsh Folk
Museum shows many similarities of construction and design to the
nineteenth-century wagons of the Vale of Glamorgan. This ox-
drawn vehicle measures as much as 10 ft. 1 in. in length and
6 ft. 5 in. broad and in its dimensions it is not much smaller than
the Glamorgan bow wagon of later times. The wain has broad

5   Cornish wain

sideboards, the sides are strengthened by numerous wooden spindles
and the floorboards are laid lengthways.[29]

During the course of this survey a vehicle which most clearly
illustrates a transitional stage between the ox-drawn cart and the
horse-drawn bow wagon was seen in Denbighshire. Although this
cart was in a derelict condition, the main features of its construction
were quite easily observed. It had a long narrow body with sides
built up from a large number of wooden spindles. The overhanging
sideboards were wide, they curved archwise over the wheels, with
elaborately forged iron side supports which held the sideboards and
sides in place. The dimensions of the cart closely follow those of a
bow wagon,[30] and it is most likely that vehicles of this type were
once common in the Celtic west.

It is interesting to note in passing that the Welsh name for a
harvest cart or wain is *men*, a word which is derived from the
Gaulish *benna*[31]—a cart of wicker, or basket work. A Welsh poet of

the sixteenth century, William Llŷn, described an ox wain in great detail in a poem *Cywydd i Ofyn Men*. The body is likened to a *cawell* (basket). Peate suggests[32] that the complicated system of wooden spindles on the eighteenth-century Glamorgan ox wain at the Welsh Folk Museum and on many bow wagons, may indeed be derived from the basket-work body of the medieval *men*, the Gaulish *benna*, and the vehicles of Spain, Portugal and Southern France, which have persisted unchanged to the present day.

The development of the English wagon, therefore, may be envisaged as follows. In post-medieval times when it became the custom to use four-wheeled vehicles on English farms for harvesting, many regions, notably in Eastern England and the Midlands, adapted the continental box wagon. Craftsmen altered the design of the box wagon to suit the needs of their own localities, and the influence of some existing vehicle, notably the box cart or tumbril, played an important part in determining the construction of the new vehicles. The box wagon may therefore be regarded as a combination of continental design with that of the medieval box cart to produce a wagon adapted to the needs of the various localities.

During medieval times in the whole of western and northern Britain there were ox-drawn harvest carts with wheelguards and arched sideboards which were probably the vehicles described as 'long carts' by Piers Plowman and as 'wains' by Fitzherbert and Tusser.

In the West Country, in Wessex, the Cotswolds, Wales and the South West Peninsula, the wagon was combined with the design of a vehicle already known there, notably the ox-drawn harvest wain, and it is almost certain that the bow wagon as such did not come into existence until at least the last quarter of the eighteenth century.

These two types of vehicle, the box wagon and the bow wagon, form the basis of this study. They represent the excellent work of village craftsmen in the late eighteenth and nineteenth centuries. Despite the fact that one wagon took as long as six months to build, it was sold for a sum that varied from as little as £20 to £40.[33] As one Oxfordshire craftsman said,

'Very often in winter when we were unable to go out to the farms to do repair work' (wheelwrights were, more often than not, carpenters as well), 'we were only too pleased to have something to do. It was during the winter months that we built most of our wagons and although we rarely made much of a profit on a wagon after a winter's work, it kept us from being idle for six months.'

Towards the end of the nineteenth century, the box and bow wagons ceased to be built, except in some isolated districts where traditional types were built up to 1914. Over the major part of the country, however, the last quarter of the nineteenth century saw the end of the spindle- and panel-sided wagons. There are two factors which contributed to this. Firstly, the growth of large scale manufacturers and the easy distribution of their products meant severe competition for the village craftsmen. Secondly, from the late-eighteenth century, Scotch carts were imported into England, and by the last quarter of the nineteenth century the Scotch cart was in use throughout the country. The few wagons built by wheelwrights at that time show very definite influence of the Scotch cart in their design. These wagons, known in some parts of the country as 'barge wagons', are plank sided, with deal planks supported by very plain iron supports (Fig. 6). These supports replaced the intricate pattern of wooden spindles and finely shaped iron and wooden side supports which characterised the older wagons. In addition the wheels of the barge wagon were made much smaller than those of the traditional wagons; the forewheels in many cases being so small as to sweep right under the body.

6   Barge wagon

7 Boat wagon

Another vehicle that came into prominence during the last decade of the nineteenth century was the so-called 'boat wagon', which derives its name from the fact that it had a very shallow body, with sides sloping outwards from the floor (Fig. 7). Usually it had leaf springs and there was no coupling pole joining the two carriages.

Although the trolley, a flat platform placed on four wheels, often of equal size, was known in the early nineteenth century[34] (Fig. 34), it only became common at the end of the century. Until recently trolleys were very widely used for harvesting in some parts of the country, especially in the West Riding of Yorkshire, the Cheshire Plain and Worcestershire.

It is interesting to note that although some of the large-scale manufacturers of agricultural vehicles made wagons, which were a completely new departure from traditional shapes and designs, some took as their pattern the older village-made vehicles. For example, a foundry in Salisbury, Wiltshire, during the first decade of the present century produced two types of vehicle—which they described as a 'Dorset' wagon and a 'Gloucestershire' wagon. Although they had the general shape of the two old regional types, these vehicles were plank sided and small wheeled, and the designs as a whole were much simplified and standardised.[35]

# NOTES

1. CHILDE, V. G.   *What Happened in History*–London 1952–p. 82
2. CLARKE, J. G. D.   *Prehistoric Europe*–London 1952–p. 304
3. CZEKANOWSKI, J.   'Z Dziejow Wozu i Zaprzegu', *Lud* (Organ of the Polish Ethnographical Society)–Cracow–Poznan 1952–p. 114–Translation from the Polish by C. Z. Sliwowski (unpublished)
4. *Ibid.*–p. 115
5. *Ibid.*–p. 115
6. CLARKE, J. G. D.   *op cit.*–p. 305
7. *Ibid.*–pp. 304-5
8. [This construction was not unknown in nineteenth-century Britain, especially on such vehicles as hearses and fire engines. A manual fire engine at the Museum of English Rural Life, Reading, from Tenterden in Kent, dating from the 1860s has a completely rigid undercarriage. Although the vehicle is 10 feet long and requires at least eight men to push it, it has no lock at all.]
9. CZEKANOWSKI, J.   *op cit.*–p. 114
10. FILIP, JAN.   'The Prehistoric carriage and the Origin of the Modern carriage', *Vestnik* (Bulletin of the Czechoslovak Agricultural Museum)–Prague 1936–p. 142
11. BERG, GOSTA.   *Sledges and Wheeled Vehicles* (*Nordiska Museets Handlingar 4*)–Stockholm and Copenhagen 1935–p. 159
12. HAUDRICOURT, A. G.   'Contribution a la Geographie et l'Ethnologie de la Voiture', *La Review de Geographie Humaine et d'Ethnologie*–Paris 1948–p. 61–Haudricourt draws on linguistic evidence to show the Central Asian origins of the horse-collar. He goes even further and suggests that the collar was invented by the Altai Turks or Mongols as a development from pack saddles used on Bactrian camels. He supports his theory by the fact that the collar, like the pack-saddle, was originally fastened to the horse's chest. Haudricourt's linguistic table of harness parts is as follows,

| *Name* | *Language* | *Meaning* | *Name* | *Language* | *Meaning* |
|---|---|---|---|---|---|
| Komo | Manchurian | pack-saddle | Qamyt | Kirghiz | horse-collar |
| Qom | Mongolian | both pack-saddle and collar | Qomyt | Tatar | horse-collar |
|  |  |  | Khamat | Chuvash | horse-collar |
|  |  |  | Khomut | Russian | horse-collar |
| Qom | Altai | both pack-saddle and collar | Khamy | Ukrainian | harness |
|  |  |  | Kamantai | Lithuanian | horse-collar |
|  |  |  | Hamut | Finnish | horse-collar |
| Qom | Kazak | pack-saddle | Chomat | Polish | horse-collar |
| Hom | Tibetan | pack-saddle | Chomout | Czech | horse-collar |
| Ham | Rumanian | harness | Homut | Serbo- |  |
| Ham | Hungarian | harness |  | croat | horse-collar |
| Ham | Croat | traces | Kummet | German | horse-collar |
| Hame | Rhine | horse-collar | Comat | Romansh | horse-collar |
| Haam | Dutch | harness | Comaco | Venetian |  |
| Hame | English |  |  | (dial.) | horse-collar |

13. CZEKANOWSKI, J.   *op. cit.*–p. 120
14. WILLARD, J. F.   The Use of Carts in the fourteenth century, *History*–Vol. 17–1932–p. 248
15. PARKES, J.   *Travel in England in the Seventeenth Century*–Oxford 1925–p. 7–'The intrusion of the Dutch form, *waggon*, into the English language

during the sixteenth century suggests that the four-wheeled waggon was probably introduced from the Low Countries at this period, the new word being borrowed to differentiate it from the Tudor two-wheeled wain hitherto employed.'

STOWE, J.  *The Annales of England*–London 1601–p. 867–'In the year 1564 Gwylliem Boonan, a Dutchman brought the use of coaches into England, and about that time began long wagons to come into use, such as now come to London.'

[R. H. Lane, in a letter, notes that a Proclamation of 20th July, 1618, decreed, 'Of the recent decay of the highways and bridges is due to common carriers, who *now* use four-wheeled wagons drawn by eight, nine or ten horses and carrying sixty or seventy hundredweights at a time, where heretofore they used two-wheeled carts carrying twenty hundredweights.']

16. FITZHERBERT, A.  *The Book of Husbandry*–Edited by W. W. Skeat from the edition of 1534–English Dialect Society, No. 37–London 1882 –p. 5

17. *Ibid.*–p. 5

18. TUSSER, T.  *His Hundred Points of Good Husbandry*–Edited by D. Hartley–London 1931–p. 91–
    'Horse, oxen, tumbril, cart wagon and wain,
    The lighter and stronger, the greater they gain.'

19. *Ibid.*–p. 76–  'Let cart be well searched without and within
    Well clouted and greased ere hay time begin.'
    *Ibid.*–p. 142–  'Strong axle treed cart that is clouted and shod
    Cart ladder and wimble with percer and pod.'

20. SEEBOHM, M. E.  *The Evolution of the English Farm*–2nd edition–London 1952–p. 132 *et seq.*

21. SCOULUODI, I.  'Alien immigrations into, and alien communities in London', *Bulletin of the Institute of Historical Research*–Vol. 16–1938-1939–p. 193–The 4700 aliens resident in London in 1567 the most important group were engaged in the clothing and textile trades, while second in importance were those engaged in transport.

22. [At the Rijksmuseum, Amsterdam.]

23. [At the British Museum.]

24. WAITING, H. R. and J. B. PASSMORE.  *The English Farm Wagon*–MS at the Science Museum dated 1936–p. 2

25. LISLE, EDWARD.  *Observations in Husbandry*–London 1757–p. 37–Wagons were occasionally used on Hampshire farms in the late seventeenth and early eighteenth centuries. The most notable feature of the English farm wagon, namely the overhanging sideboards, was not found on the wagons that Lisle described. He does however mention the custom of tying 'the sideboards to the rathes of the wagon with leather thongs greased,' suggesting that these vehicles were adaptations of the large carrier's wagon.

26. WORGAN, G. B.  *General View of the Agriculture . . . of Cornwall*–London 1815–p. 37

27. *Ibid.*–p. 38

28. *Ibid.*–p. 57

29. PEATE, I. C.  'Some Aspects of Agricultural Transport in Wales', *Archaeologia Cambrensis*–1935–p. 227–'The wain illustrates the development of a well known Mediterranean type for Welsh harvest usage.' [The principal dimensions of the Glamorgan ox wain are, Diameter of wheels (calculated) = 72 inches, Maximum length of body = 121 inches, Maximum depth of body = 34 inches, Maximum width of body = 77 inches, Width of sideboard (front) = 14½ inches.]

30. [The following table shows the measurements of the cart seen near Llangollen in Denbighshire compared with those of a Cotswold bow wagon.

|  | | *Denbighshire*<br>*Cart* | *Cotswold Wago n* | |
| --- | --- | --- | --- | --- |
| *WHEELS* | | | *Rear* | *Fore* |
| Diameter | . . . | 55 inches | 58 inches | 48 inches |
| Number of spokes | . . | 12 | 12 | 10 |
| Width of Tyre | . . | 4 inches | 2½ inches | 2½ inches |
| Track of Wheels | . . | 67 inches | 66 inches | 66 inches |
| *FRONT* | | | | |
| Maximum Width (Top) | . | 79 inches | 75 inches | |
| Depth of Frontboard | . | 9 inches | 20 inches | |
| Width of Sideboard | . | 16 inches | 14 inches | |
| Width of Forebridge | . | 50 inches | 45 inches | |
| *BODY* | | | | |
| Total Length (Top) | . | 147 inches | 158 inches | |
| Front (Top) to ground | . | 53 inches (app.) | 65 inches | |
| Bow to Ground | . . | 56 inches (app.) | 64 inches | |
| Back (Top) to Ground | . | 53 inches (app.) | 58 inches | |

31. PEATE, I. C.  *op. cit.*–p. 229
32. *Ibid.*–p. 229
33. [A few examples of wagon prices in the late eighteenth and nineteenth centuries are listed below. They were extracted from a variety of sources including County Agricultural Surveys and wheelwrights' accounts.

| *County* | | | | *Year* | *Average Price per Wagon* |
| --- | --- | --- | --- | --- | --- |
| Bedford | . | . | . | 1813 | £33 12s |
| Warwick | . | . | . | 1815 | £45 (4-horse type) |
| Somerset | . | . | . | 1798 | £26 (narrow wheeled) |
| | | | | | £36 (broad wheeled) |
| Gloucester (Cotswold) | . | . | | 1789 | £20 |
| Gloucester (Cotswold) | . | . | | 1880 | £28 |
| Gloucester (Vale) | . | . | | 1880 | £34 (+ £2 2s for |
| | | | | | double shafts) |
| Dorset | . | . | . | 1880 | £20 |
| Sussex | . | . | . | 1870 | £30 |
| Shropshire | . | . | . | 1880 | £37 |
| Oxford | . | . | . | 1880 | £28 |
| Lincolnshire | . | . | . | 1890 | £29 |
| Suffolk | . | . | . | 1880 | £36 |
| Wiltshire | . | . | . | 1860 | £25 |
| Kent | . | . | . | 1880 | £30 |
| Northampton | . | . | . | 1870 | £33 10s |
| Hereford | . | . | . | 1870 | £36 |
| Glamorgan | . | . | . | 1880 | £32 10s |

PITT, W.  *General View of the Agriculture* . . . *of Worcester*–London–1813– p. 48

34. [The following table gives the measurements of three early twentieth-century mass-produced wagons. The plank-sided wagon was seen at Powick, Worcestershire; the boat wagon was built by a Basingstoke firm

and was seen at Romsey. The trolley was seen near Nantwich in Cheshire.

| | Plank Sided Barge Wagon Worcester 1902 | | Boat Wagon Hampshire 1904 | | Trolley Cheshire c. 1904 | |
|---|---|---|---|---|---|---|
| **WHEELS** | | | | | | |
| Diameter | 47″ | 39″ | 48″ | 36″ | 42″ | 42″ |
| Method of Tyring | Hoop | Hoop | Hoop | Hoop | Hoop | Hoop |
| Width of Tyre | 2½″ | 2½″ | 3″ | 3″ | 2″ | 2″ |
| Track of Wheels | 64″ | 64″ | 60″ | 60″ | 54″ | 54″ |
| **FRONT** | | | | | | |
| Maximum Width (top) | 73″ | | 78″ | | 70″ | |
| Depth of Frontboard | 17½″ | | 18″ | | — | |
| Width of Sideboard | 4″ | | — | | — | |
| Width of Forebridge | 50″ | | 64″ | | — | |
| **BODY** | | | | | | |
| Total Length (top) | 117″ | | 132″ | | 144″ | |
| Total Length (bottom) | 109″ | | 130″ | | — | |
| Front (top) to ground | 38″ | | 57″ | | 48″ | |
| Mid Part (top) to ground | 42″ | | 51″ | | 48″ | |
| Rear (top) to ground | 44″ | | 57″ | | 48″ | |
| Front (bottom) to ground | 31″ | | 44″ | | — | |
| Mid Point (bottom) to ground | 35″ | | 36″ | | — | |
| Rear (bottom) to ground | 37″ | | 42″ | | — | |
| **COLOURS** | Yellow | | Blue | | Brown | |
| **LOCK** | Full | | Full | | Full | |

# CONSTRUCTION

## II

### 1 WHEELS

The earliest form of wheel was undoubtedly the solid disc, which was invented by the inhabitants of the Tigris-Euphrates Valley soon after 3500 B.C.[1] The technique of wheel construction spread gradually from the Mesopotamian centre to all the known world, reaching the Indus Valley and the steppes of Central Asia by 2500 B.C. By 2000 B.C. wheeled vehicles were known in Crete and by 1500 B.C. they were known in Egypt, Greece and Georgia. Some time before 1000 B.C. the technique of building disc wheels reached Northern Italy and then spread through the Alpine passes to Central and Western Europe, arriving in the island of Britain by the fifth century B.C.[2] There are early models and sculptures of two-wheeled and four-wheeled vehicles as well as actual examples preserved in the tombs of the third millenium in the Near and Middle East. All these vehicles have disc wheels bound with leather thonging, attached to the rims by copper nails. In all cases early wheels are built of three planks morticed together and rounded. Nowhere in the Mediterranean area and South West Asia are wheels found carved from one solid piece of wood. The absence of one-piece discs in this vast region may be explained by the fact that trees large enough to yield solid planks of adequate width were rare. It is only in well-wooded Denmark, and then at a much later date, that one-piece wheels occur side by side with tripartite discs.[3]

Early wheels generally have a small diameter. For example, a wheel of 2750 B.C. from Kish is no more than 19·6 inches across; another of 200 B.C. from Dystrup Mose in Denmark has a diameter of only 11·2 inches. The treads are very narrow, a wheel from Susa, for example, having a width of no more than 1·5 inches.[4]

Although the disc was the earliest form of wheel, the first vehicles, chariots and cult vehicles, actually represented in the archaeological record north of the Alps, already had spokes. Disc wheels were adequate for a vehicle carrying heavy goods, but less satisfactory on vehicles where speed and manoeuvrability were essential. At a very early date, therefore, spoked wheels replaced disc wheels on certain special types of vehicle and it seems probable that both types were diffused from a Mesopotamia centre at approximately the same date.

The lighter-spoked vehicles occur first around 2000 B.C. when they are depicted on clay models from Chagar Bazar in Northern Mesopotamia. A chariot carved in a chieftain's tomb at Kivik in South East Sweden is dated around 1000 B.C.[5], and by 500 B.C. the technique of building both disc and spoked wheels was known to people from China to Britain.

It is interesting to note that even up to the Middle Ages the use of spokes was limited to vehicles of war and ceremony, and that disc wheels continued to be used for carts and wagons carrying bulky goods. On continental peripheries, solid-wheeled carts remained in vogue right down to the present century, and may still be seen in remote areas. For example, disc-wheeled vehicles have only recently disappeared from West and North Sweden, Scotland and Wales, while they are still used occasionally in Norway, the Iberian Peninsula and Asia Minor. They may still be seen in Ireland, and Estyn Evans[6] suggests that the disc wheel has had a considerable influence on the shape of the spoked wheel in that country. The spoked wheel, like the disc, has a very narrow rim, the diameter is small and the spokes have little dish.

From the foregoing it will be apparent that the origin of the spoked wheel is shrouded in mystery, and archaeological evidence of its evolution is limited. Various authors have, however, put forward theories about its evolution. Motefindt,[7] for example, thinks that the multiplication of openings carved in the planks of a disc wheel would naturally lead to spokes in time. Crescent-shaped slits were carved around the hub of a few early wheels from Northern Italy and Denmark, similar to wheels in modern Sind. But these slits bear little relation to a radial arrangement of spokes. It seems unlikely that this type of wheel could be the predecessor of the spoked variety. Childe[8] notes that the only wheel which might possibly mark a transitional stage between the tripartite disc and the true spoked wheel, is one from the lake dwelling of Mercurago in Northern Italy. This has a solid plank carved to form a hub and two spokes. Into these spokes two others have been morticed at right angles. Even so, this wheel which is dated around 1000 B.C. considerably post-dates fine specimens of true spoked wheels excavated in the Near East.

It could be that the conception of a spoked wheel was completely new, and not a modification of the tripartite disc. The stimulus was probably a desire to improve speed and manoeuvrability by reducing the weight of vehicles. While the disc was strong and durable, it was far too heavy and clumsy for the wheels of war chariots where pace and easy control were essential. Discs were simple to construct, hence their long survival in some parts of the world; spoked wheels were much more costly and demanded more labour and greater skill. For these reasons spoked wheels, until relatively recent times, were limited to the elaborate vehicles of ceremony and war, where cost and labour were not the prime considerations.

While the earliest spoked wheel is represented by painted clay models of 2000 B.C. from Chagar Bazar, other wheels of the same date have been found in Cappadocia (Central Turkey) and at Hissa in north-east Persia. They were adopted in Egypt soon after 1600 B.C. and are repeatedly depicted in Crete and Mycenae by 1500 B.C.[9] In China, too, the wheelwrights' art was very highly developed as early as the second century B.C. and the *Chou Li*[10] describes in detail the technique of building a multi-spoked wheel, a technique that differed only slightly from that adopted by modern wheelwrights in Britain.

Gradually the wheelwright's art spread to all parts of the known world. In Europe it reached its highest technical level among the Celtic inhabitants of the Rhineland and Bohemia, who later introduced the art to other parts of North Europe and Britain. By 500 B.C. Celtic craftsmen were shaping the rim or felloe of a wheel from a single piece of ash steamed or heated to shape. The ends were bevelled, overlapped and held in place by a metal swathe. This ingenious technique of manufacture was unknown in the Mediterranean region but according to Childe[11] it was referred to 'in one of the Vedic Hymns chanted by the Aryans who invaded India between 2000 and 1000 B.C.' and it was probably brought to Western Europe by the almost mythical Indo-European people.

The details of early spoked wheels are derived from hearses and chariots, many of them excavated from the burial places of tribal chieftains. All reflect a high standard of workmanship and uniformity of construction which suggests the existence of a well-established tradition in Iron Age Europe.

Among the most important features of these early wheels is the large number of spokes and the construction of felloes from a single piece of wood, steamed into shape and held by a metal tyre. The tyre was heated and then shrunk on to the rim. The spoke tongues were generally short and the hubs long and small in diameter.

The Dejbjerg cult wagon shows the ability of European wheelwrights around 100 B.C. Each of the four wheels had fourteen spokes and each felloe consisted of a single piece of wood held by a metal swathe. The wheels were shod with iron tyres.

Wheels and fragments of wheels dating from the Iron Age have also been found in many parts of Britain. From the Glastonbury Lake Village, for example,[12] a turned ash hub with twelve circular spoke motices was excavated. The centre of this hub was hollowed to accommodate an axle arm $3\frac{1}{2}$ inches in diameter. The hub itself measured $13\frac{7}{8}$ inches long and had an exterior diameter of $5\frac{7}{16}$ inches at the ends, with a central diameter of $7\frac{7}{8}$ inches. The shaft of a spoke from the same site was $11\frac{9}{16}$ inches long so that the diameter of the whole wheel was $31\frac{1}{8}$ inches. The tongues of the spokes were short, being no more than $\frac{9}{16}$ inch long. These discoveries at Glastonbury suggest that by the end of the prehistoric period, at least in the area of La Tène culture, the wheelwright was an established member of the village community.

From the Roman frontier post at Newstead another wheel was found, presumably made by native craftsmen.[13] This had an outside diameter of 36 inches. The lathe-turned elm hub was $15\frac{1}{2}$ inches long with a diameter of $8\frac{1}{2}$ inches at the centre and $5\frac{7}{8}$ inches at the ends. It had willow spokes fitted into the felloe with round tongues and into the hub with square tenons. The felloe itself was shaped from a single piece of ash, the ends were bolted together by a metal plate, and the whole wheel was bound with a tyre $\frac{3}{8}$ inch thick and $1\frac{3}{4}$ inches wide. A wheel with a diameter of 41 inches was excavated from the same site.[14] This larger wheel was similar to that used on the English farm wagon during the last hundred years. It had six wooden felloes dowelled together. Two, almost square, spokes passed into each felloe.

Before the arrival of the Romans in Britain, the Iron Age inhabitants of the country knew all the techniques of wheelwrighting, techniques that were to remain basically unchanged for the next two thousand years. At this early period, hubs were lathe turned and spokes had square tenons and round tongues which passed through the felloes. The technique of dowelling felloes and

C

fitting two spokes to each, the method of boxing the wheel and shrinking a heated tyre on to the felloes, were all well understood by the inhabitants of Western Europe.

Little is known of the structure of wheeled vehicles between the fall of the Roman Empire and the late Middle Ages, but it is unlikely that wheels showed any remarkable technical change. It

8    Wheel from the Roman frontier post at Newstead

seems, however, that craftsmanship degenerated considerably during the Dark Ages and the early medieval period. Wheels of early medieval date which have been excavated from the Scandinavian bogs, differ little in essentials from those of the Iron Age but, in detailed construction, are definitely inferior. A wheel found in a bog at Södermanland, for example, is equipped with six deep roughly-shaped felloes with one spoke to each.

The next brief glimpse of wheel construction is in a calendar picture from an eleventh-century manuscript. This shows a two-wheeled cart with each wheel made up of six deep felloes, one spoke to each. The wheels are unshod, a common feature on medieval carts as iron-bound wheels were often forbidden in towns because of their effect on road surfaces.[15]

In the fourteenth century Luttrell Psalter two vehicles are illustrated, a harvest cart and a travelling wagon. Each wheel of the cart is shown with six spokes and six felloes, the latter in this case being much shallower than those of the eleventh-century vehicle. The felloe junctions are lagged with pieces of metal, while the rim itself is shod with a number of lugs or nails to prevent slipping. The other vehicle illustrated in the Psalter is a highly ornamented travelling coach. Its wheels have six spokes and a narrow one-piece felloe shod with an iron hoop.

Evidence is scanty, but it seems that in the late Middle Ages the better type of spoked wheel with narrow felloes and tyres shrunk on was limited to cult vehicles and the travelling wagons of the rich. Agricultural vehicles, on the other hand, were either equipped with disc wheels or with roughly shaped spoke wheels. It seems doubtful whether the full potentiality of the spoked wheel for sustaining heavy loads was realised before the Late Middle Ages.[16]

With the general improvement of transport in the sixteenth century, spoked wheels of excellent design and durability became common, especially on the growing number of road wagons that carried merchandise all over the country. The large wheels of these vehicles were either shod with hoops or with crescent-shaped pieces of iron called strakes. In addition, attempts were made to reduce the weight of the wheels by cutting away all superfluous timber, especially at the back of the spokes where a great thickness of wood was not required.

By far the most important innovation of the sixteenth century was a new form of spoked wheel, the dished wheel (Fig. 9). Prior to the sixteenth century, all European wheels represented sections of cylinders, with the spokes leaving the nave at right angles. The dished wheel, on the other hand, is a shallow cone with each spoke slanting outwards from the nave. When hung on an axle arm the lowest spoke of the wheel, which momentarily takes the weight of a vehicle, is in a perpendicular position, or is at least in a position very near the vertical. The upper part of the wheel overhangs to a point nearly above the end of the nave. To produce this effect the axle arms have to be bent downwards towards the ground, bringing the lowest spoke near to the perpendicular. That part of the wheel in contact with the ground has to lie in a horizontal plane, so that the rim surface formed by the outer edge of the felloe has to be at right angles to the line of the corresponding spoke. In a dished wheel this surface is cone shaped, and the tyres also must be coned to fit the rim.

Neither writers nor craftsmen have been able to decide satis-

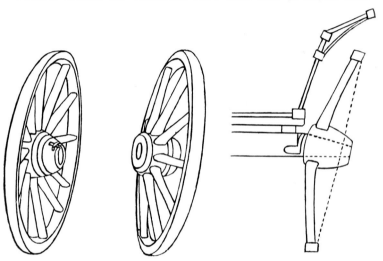

9   The dished wheel (after Sturt)

factorily whether this new, shallow coned wheel is superior to its cylindrical predecessor. Some say that dish is absolutely essential in making a wheel that is both durable and strong, others say dish is quite unnecessary; no one has ever given ample reasons why it should be found on most cart and wagon wheels between the sixteenth and nineteenth centuries.

Surprisingly enough the technique of dishing wheels was well known to Chinese wheelwrights from at least the second century B.C.,[17] but despite its antiquity in Asia, there is no evidence that the technique was known in Europe before A.D. 1500. Jost Ammen's woodcut of a wheelwright's shop in A.D. 1568 provides the first concrete evidence for the practice of dishing in Europe,[18] but after that date the dished wheel became almost universal. It is possible, of course, that wheels were first dished unintentionally and not for any technical reason. This is a very easy thing to do, since the iron tyre is forced over the wheel when hot and then it shrinks. 'As the tyre contracts the spokes are pushed into the hub with terrific force and unless the wheel is held down the hub will rise above the level of the felloes, thus turning the wheel into a shallow cone.'[19]

There have been many advocates of dish, each writer going into considerable detail to point out the constructional superiority of the dished wheel. James Small, for example,[20] condemns the cylindrical wheel on the grounds that it takes up a great deal of mud which falls on the nave and penetrates through to the axle arm causing rapid wear. By dishing the wheel, says Small, the mud falls well clear of the nave, so that a vehicle with dished wheels has a much longer life than that with cylindrical wheels. Small's reasoning is not really valid; unless a wheel is so greatly dished that the top leans well beyond the nave, the mud will fall on the hub and in time penetrate into the axle arm just as if the wheel were totally undished.

Small, George Sturt,[21] and many others have advocated dish on the grounds that the dished wheel is able to bear the lateral thrust caused by the side-to-side movement of a moving vehicle better than the cylindrical wheel. The jolting motion of a vehicle causes an undished wheel to bear hard on the inner side of the axle arm, so that the spoke tenons in the nave are greatly strained. By dishing, which sets the spokes against the jolts of lateral movement, this weakness is alleviated. Although dishing may to a certain extent contribute to strength, this explanation fails to show why some wheels are dished more than others and some hardly dished at all. Many Dorset wagons are equipped with completely undished wheels, while mass-produced boat and barge wagons are always equipped with undished wheels of a small diameter. Yet these vehicles have stood up to many years of work on deeply-rutted farm tracks without any apparent weakening of spokes and axles. Experienced wheelwrights shake their heads at Sturt's explanation for, they say, although there is lateral movement on a moving vehicle, it has very little effect on a large four-wheeled wagon.

Another advocate of dish was John Rickman (1771-1840)[22] who pointed out the technical advantages of the dished wheel over the cylindrical type. He notes that undished wheels are 'awkward in turning and do not possess the advantage resulting from the outward slope of the upper part of the felloe of a dished wheel, whereby in mutual collision with another carriage, the conflicting wheels yield gradually, and prevent any violent shock of the naves, thus protected. Dished wheels also possess a greater strength of construction, for the felloe being compressible owing to the slant of the spokes which support it, is strained tight by the contraction of the iron tilt in covering, and being thereby kept in a state of tension, is not liable to become loose in the joints, and those wheels are capable of resisting shocks upon a rough roadway. A wheel falling into a cavity or rut receives from the sloping load, not only an extra pressure, but also a lateral thrust; the tendency of which is to force the nave outwards through the felloe; but the nave of a dished wheel resembles the crown of an arch, bearing upon the felloe as a base, which cannot be extended, because being of a circular figure, it admits of no extension while entire, thus the felloe is equivalent of the abutments of this arch, or rather to the tie beam of a trussed frame.'

The practice of dishing wheels is not limited to the British Isles, but oddly enough it is far more pronounced on British vehicles than on any others. This may well be due to legislation; although Parliamentary Acts may not have been prime factors in determining dish, they could well explain the very pronounced dish on some of the older wagon wheels. By an Act of 1773, it was said that no carriage using broad wheels should make a track more than 68 inches wide. Although this fixed the distance that wagon wheels were to be apart at the bottom, it did not mention the top of the wheels. The farmers, carriers and wheelwrights overcame this restriction by dishing their wheels very greatly so that they conformed to the legal distance at the bottom but were wide apart at the top. This left room for a capacious body, another argument in support of dish. Although the Act of 1773 related specifically to broad-wheeled vehicles, the narrow-wheeled type would also tend to be the same width because traction was easy on an already flattened track.

While some writers advocated dish, others condemned it in no uncertain terms, regarding it as 'A monstrosity beneath the contempt of modern mechanisms.'[23] Government reports objected to it mainly because dished wheels broke up road surfaces much quicker than the cylindrical type. An Act of 1822 concerned with the preservation of road surfaces, forbids dished wheels and states that 'in each pair of wheels . . . the lower parts when resting on the ground shall be at the same distance from each other as the upper parts of such pairs of wheels'. This proviso, which was later repealed in 1835, prevented the use of dished wheels in any form.

An earlier Report of 1809 is equally insistent in its objections to dished wheels and, indeed, even goes as far as to say that the dished wheel was structurally inferior to the cylindrical type. The only advantage the conical wheel has over the cylindrical, says this Report, is that it allows a wide body. The main objection to the dished wheel was, of course, its tendency to break up the surface of the road. The Report mentions the custom of treble-tyring wheels, and notes that by an Act of 1776 (16 Geo. III), the central tyre was allowed to project one inch more than the other two. On hard roads such wheels would run on a single tyre iron which would bear the whole weight of the load. In effect, conical wheels on hard roads would resemble a narrow-tyred vehicle, rather than the broad-wheeled vehicle which the law demanded. The observations of the committee on the respective merits of cylindrical and conical wheels were as follows:

'(1) That when the wheels are very narrow, then there is little difference in the power required to draw the same load.

(2) In conical wheels, the power required to draw the same load is considerably increased by increasing the breadth of wheels, and that all the increase of the labour of the cattle is applied to the destruction of the road.

(3) On cylindrical wheels, the same power draws the carriage upon smooth roads with equal ease, whether the wheels be broad or narrow, but by the use of such broad wheels, the roads, instead of being destroyed, are consolidated and improved.

(4) That great fluctuations take place in the power necessary to draw the same load on conical wheels, according to the circumstances of the wheel bearing the narrow parts of its rim.

(5) No such difficulty of resistance happens under the same circumstances with the cylindrical broad wheel.

(6) From every circumstance the cylindrical wheel is preferable to the conical in every state of the roads, and in whatever state they may be the cylindrical improves and the conical impairs them.

(7) . . . As regards the preservation of the roads and labour

of the cattle, the cylindrical shape of the wheel is preferable to any other possible shape, it being the only one that has the same velocity at every part of its rim, and that has no dragging or rubbing nor any tendency to grind or derange the materials, nor to leave the surface of the road in a condition to imbibe or to admit water.'

And so the argument for and against dish has raged for nearly three centuries, but whether dish is 'a monstrosity beneath the contempt of modern mechanisms' or whether 'a wheel lacking it could not be trusted to travel a mile safely' are two sides of an argument which has never been settled.[24]

The design of wheels changed little between the sixteenth and mid-nineteenth centuries and the spoked, dished and tyred wheel was widely used throughout Britain. On two wheeled farm vehicles, however, disc and untyred spoked wheels were widely used until the close of the eighteenth century. At that time with the appearance of the farm wagon, dished and tyred vehicles gradually replaced the older types.

For road wagons, on the other hand, the size and detailed construction of wheels was largely determined by Acts of Parliament during the sixteenth, seventeenth and eighteenth centuries. Within these centuries successive governments tried to improve the condition of English roads by passing numerous Acts which aimed to limit the size of the wheels, their shape, the type of nails used for fastening strakes and the position of the wheels. This policy, pursued for over two centuries, sought to adapt the traffic to the indifferent roads of the time rather than adapt the roads to the traffic. Despite this spate of Acts which were formulated and repealed at frequent intervals, the construction of wheels changed little between 1550 and 1850.

In the mid-nineteenth century, when the coach builder's art had reached a high level of achievement, village wheelwrights took over the technique of the coach builder for the construction of farm carts and wagons. It became the practice to build much smaller and narrower wheels;[25] dish became far less pronounced, and among the new techniques the practice of staggering the spokes became common.

## 2 UNDERCARRIAGE

The earliest form of axle was a square or cylindrical block of wood joining a pair of tripartite disc wheels. As the vehicle moved, the axle revolved with the wheels. For this reason some form of fixture was necessary to ensure that the axle and wheels did not slip away from the body as the vehicle moved. Wherever disc wheels have persisted, so too have the revolving axle and the various methods of attaching the axle to the body. In ancient Mesopotamia, for example, a series of leather thongs nailed to the underside of the body went round the axle and kept it in place.[26] This method of construction is basically the same as that adopted in northern Europe in modern times. Carts in early nineteenth-century Scotland, for example, were equipped with two semi-circular hoops of birch which kept the undercarriage in place.[27]

Other methods of attaching axles to cart bodies have been known from very early times; perhaps the most widespread was that of fitting a number of pairs of wooden wedges to the underside of the wagon body. This method may still be seen on many vehicles in isolated regions; indeed, in Cantabria the almost human sound produced by the friction of an axle revolving against wedges has given one cart the rather romantic name of *carro que canta* (groaning or singing cart).[28]

Although revolving axles have been universal from prehistoric to modern times, they are never found on four-wheeled vehicles. Even the earliest wagons were equipped with spoked wheels and stationary axles, whereas throughout the centuries carts have had revolving axles. The explanation must be that the wagon in prehistoric and protohistoric times was specifically a ceremonial vehicle displaying only the most expensive and elaborate methods of construction.

Despite the survival of the revolving axle in some parts of the world, a form of wheel bearing has been known from a very early date. The Dejbjerg wagon had a series of roller bearings within the nave, and a straight, untapered wooden axle.[29] From Iron Age Glastonbury, too,[30] an axle cut from a solid piece of oak was excavated. This was circular in cross-section and the arm had an untapered diameter of $3\frac{1}{2}$ inches. All other axles from prehistoric and early medieval excavations are of this cylindrical wooden armed type and it is not until the Late Middle Ages that tapered axle arms occur. The factor that led to a change in axle design was the appearance of dish. As long as wheels remained undished, the untapered axle could be used, because the axle arm naturally enters the nave at right angles to the spoke. However, when wheels were dished, the axle arm had to be given a distinct downward pitch to place it at right angles to the lower spoke of the dished wheel. The arms were also tapered and shod with countersunk plates of iron to take the wear of the revolving wheel.

On every English wagon since the sixteenth century all-wooden axles were fitted and continued in vogue for as long as village craftsmen built four-wheeled vehicles. In some regions iron axle arms were fitted to wooden beds from the early 1800s, while around 1840 in other regions these were replaced by all-iron axles.

In addition to axles, the undercarriage of a wagon consists of a movable forecarriage and a coupling pole which links the rear axles and wheels with the fore-part of the vehicle. Early wagons were built from two triangular carts and the earliest type was characterised by a fixed forecarriage; with the substitution of horse for ox traction, and the use of side-shafts instead of a draught pole, a radical change in forecarriage design occurred. Instead of fixed fore-axle and carriage, a movable and independently constructed forecarriage was introduced. As far as can be ascertained, the construction of this carriage changed little over the centuries, from post-Roman times when the horse traction replaced ox draught.

In the same way the construction of the coupling pole has remained unchanged. This forked rear carriage may well represent the draught pole of a triangular ox cart in a vehicle which is basically a combination of two such carts.

# 3 DRAUGHT POLE, SHAFTS AND TRACES

The draught pole, or neb, was the earliest method of attaching a vehicle to draught animals, and a central pole fixed to the under-carriage of a cart or wagon passed between a pair of animals yoked to it. This method of harnessing is known as 'paired draught', for it cannot be used unless a pair of animals in line abreast are harnessed to it.

As has been said, the earliest form of wagon came about through the fusion of two triangular-framed ox carts, so that the central draught pole in such a vehicle was, in reality, an extension of the apex of the triangle formed by the foremost cart in a composite vehicle. The representation of a horse-drawn vehicle on a sherd of pottery from Odenburg in Hungary shows this construction in Early Iron Age times.[29] On the other hand, the Dejbjerg wagon shows a slightly more advanced method of construction; the draught pole is not an extension of the undercarriage but an independent construction with a forked end attached by ropes to the fore-axle.[30] In early times the draught pole was invariably associated with an immovable, non-locking forecarriage, and the locking forecarriage did not appear until the end of the Roman era.[31]

At the present time the distribution of paired draught bears a close relationship to that of ox-drawn vehicles. Where oxen were used until recent times the draught pole is still commonly found, even though oxen may not be used for draught purposes today.

As far as Great Britain is concerned, the draught pole was widely used until recently in the west and north regions where ox carts of Mediterranean origin were used. In Wales, ox wains were well known,[32] while the wains and wagons of Cornwall were in the early nineteenth century fitted either with side shafts or draught poles.[33] In East Yorkshire, too, ox-drawn wains and wagons remained common until a late period and these vehicles were 'furnished with poles similar to that of a coach'.[34] Indeed many East Riding wagons built as recently as 1920 were still equipped with central draught poles and not with shafts, although of course horses were used for draught purposes. Even when a Yorkshire wagon is equipped with side shafts, vestiges of the old traditional draught pole are apparent on the forecarriage. Parallel pieces of timber, known as hounds, converge at the front and project some distance beyond the forecarriage, so that the end of a draught pole could be fitted there, and used as an alternative to side shafts.

By the close of the Roman era, side shafts had made an appearance in temperate Europe and these gradually replaced draught poles on many wagons. A key to the evolution of wagon shafts may be found in a series of vehicles which have persisted in some of the more isolated parts of Europe.

The simplest arrangement is that found on some vehicles in Scandinavia, where wagons have draught poles attached to the forecarriage and also straight poles attached to the outside edge of each front wheel nave. These side shafts run parallel to the central pole.[35] A slightly more advanced method of shaft construction also occurs in this region. Two central poles are attached to the forecarriage as well as the two side pieces which join the ends of the naves.[36] In these two simple arrangements the shafts are as straight as poles roughly shaped from the trunks of small trees can be.

In the third evolutionary stage represented by vehicles from both Scandinavia and western Germany,[37] the side poles attached to the naves are replaced by two roughly-hewn side-pieces joined to the two central poles by a pair of cross-pieces. The actual cross-pieces and not the central poles are pinned to the forecarriage.

The development of wagon shafts is completed by refinements in shape which allow for the curvature of the horse's body.

Most English farm wagons, apart from a few in Yorkshire and South-West England, are equipped with side shafts; and in all parts of Western Europe they are the most common method of attaching draught animals to vehicles.

In Eastern Europe, although wagons are occasionally equipped with side shafts, chain traces are usually used for harnessing. This method was probably evolved in the Middle Ages in an attempt to improve paired draught, for the chain or rope traces are often combined with a central draught pole. In the Baltic States, each wagon has a central draught pole, but the pair of horses harnessed to the vehicle are also equipped with chain traces attached to a pair of whipple trees. These whipple trees are on a swivelling cross-piece, fitted to the central pole and at right angles to it. As an improvement on central pole draught, the chain trace method is found in many parts of the European continent; it is the usual method in Eastern Europe, and is occasionally found on wagons in Western Europe. Before 1880 many English road wagons were equipped with chain traces and a central pole and they are found on brewer's drays even at the present time. Outside Europe, the famous Conestoga wagon and others used in the development of the American West and the huge Boer wagons of South Africa were all equipped both with central poles and chain traces, to which anything up to a dozen horses were harnessed.

The evolution of draught may be summarised as follows. The earliest method of attaching animals to a vehicle was with a central draught pole. This has persisted in Mediterranean Europe and other areas where draught oxen were used until recently. Towards the close of the Roman era side shafts were introduced into Europe from Central Asia, and replaced the central draught pole on many vehicles. Side shafts are used on most English farm wagons and are the most common method of harnessing in Western Europe. At the same time an attempt was made to improve paired draught by equipping wagons with chain traces. This practice has persisted in Eastern and Central Europe and is frequent in Western Europe, North America and South Africa. It was also common on English road wagons before 1880 and may be found on coaches, hearses, drays and other non-agricultural vehicles in Britain.

## 4 WAGON BODIES

Most wagons on the European continent from the Urals to the North Sea are narrow bodied and simply constructed. In parts of France, for example, the side frames of a wagon are no more than two feet apart and the floor is made by sliding a number of planks into the angle formed by the side frames and the two end frames morticed to them. The sides are constructed of a number of round wooden spindles morticed to the side-frames and sloping outwards over the low wheels to a single top-rail. When seen from the front, the wagon is almost V-shaped, and it is very cheaply and easily built.

In addition to its frequent occurrence on the European continent at the present time, the spindle-sided open-work wagon was common in medieval Britain where it was used for carrying such goods as bales of wool and loads of hay or corn. Illustrations from psalters show carts with an open framework of spindles, interesting survivals of this early form of construction being the so-called harvest wagons of East Anglia, Huntingdonshire and Cambridgeshire. A wagon seen at St. Neots, for example, was planked over the lower half of the sides, while the top half was a grille of spindles with no planking. Although the open-sided wagon may be perfectly efficient for carrying material in sacks or bundles, it cannot be used for carrying goods in bulk, such as loads of gravel, sand and loose grain. Since wagons were introduced into Britain it has been customary to plank the sides, while retaining the wooden spindles outside those planks. All the road wagons of the sixteenth, seventeenth and eighteenth centuries were spindle sided and this was continued on many regional types of farm wagon until the present century.

The second variety of side construction is panelling. In panel-sided vehicles the numerous round spindles are replaced by flat slats or ribs nailed directly to the side planks of the wagon. Sometimes these slats are thick and their edges chamfered to a delicate line. Each is morticed to the side frame and top-rail of the vehicle, so that the wagon has the appearance of being panelled. This form of construction is rare outside Britain, although occasionally found on later wagons along the coastal periphery of Western Europe.

Side panelling probably originated in the early eighteenth century as an adaptation of the technique used on the traditional two-wheeled tumbril of East Anglia and adjacent regions. These carts, common in seventeenth and eighteenth-century Britain, were always panel sided. In the eighteenth century, when it became customary to use wagons for farm transport, these were built on the same lines as the earlier carts. Just as the spindle-sided wagon owed a great deal to the design of the open-spindled medieval cart, so the panel-sided wagon owed a great deal to the design of the tumbril.

During the last quarter of the nineteenth century, a radical change occurred in the design and construction of four-wheeled wagons in Britain. Wheels became smaller, so that the fore-wheels could lock right under the wagon body; the coupling pole between the forecarriage and rear carriage disappeared to give independent suspension to the two carriages, while the spindles or standards that gave added strength to the sides of the body were superseded by two or more bare planks running the length of the wagon, supported only by a few plain iron brackets.

A number of factors were responsible for this radical change in design, the most important being the entry of large-scale manufacturers into the wagon building trade. The market was flooded by standardised vehicles that could be built cheaply and quickly. Of the few village wagon builders that survived this fierce competition, many adopted new techniques of manufacture. Another factor was the introduction of the multi-purpose Scotch cart into England. Whereas the old carts had spindle or panelled sides, their Scottish successors had plain, planked sides, and were far more cheaply constructed. The simple techniques of Scotch cart construction were adapted by English wheelwrights for the much larger four-wheeled wagons, and plank sided vehicles rapidly replaced traditional types in some parts of the country.

Despite the general change in design, the plank-sided wagon in some parts of the country still displayed characteristics of its spindle or panel-sided predecessors. In Hereford, for example, where plank-sided wagons did not make an appearance until the last decade of the nineteenth century, the older wagons were deep bodied vehicles with panelled sides and two mid-rails. Their successors, the plank-sided vehicles, were still deep bodied but, instead of the two mid-rails, two narrow deep grooves were cut from end to end of each side and lined in black, red or white. The remainder of the wagon was blue. The new Hereford wagon had almost exactly the same measurements as its panel-sided predecessor. Thus in Hereford, as in some other counties, there was a continuity of tradition in many features of wagon design. This continuity was only destroyed by the appearance of mass-produced standardised boat wagons built by large scale manufacturers in the early twentieth century.

In north-east Yorkshire, however, plank-sided wagons were not an innovation of the last quarter of the nineteenth century for they were known in that area as far back as 1800. The junctions between the planks of the wagon sides were lined in white, in sharp contrast to the remainder of the body which was brown. The early appearance of planking in Yorkshire vehicles was probably due to the influence of Scottish agricultural techniques. For example, in 1796, John Tuke[38] mentions Scotch carts being in use in Cleveland and the Vale of Mowbray.

Plank-sided wagons were introduced into the eastern counties long before they became common in the west. In Lincolnshire and in East Anglia they were well known in the 1870s, but in Hereford they were not introduced until the turn of the century. In many areas, notably Oxfordshire, the Cotswolds, the South-West Peninsula and Shropshire, they never superseded the traditional local types. It is almost certain that where the Scotch cart made an early appearance, plank-sided wagons rapidly replaced the traditional types. The Scotch cart was popularised mainly in East Anglia and Lincolnshire whence the design spread throughout the East Midlands of England. Arthur Young[39] mentions the fact that two-wheeled carts were imported by sea to Essex from Edinburgh in the early nineteenth century. At a later date the widespread settlement of Scottish farmers in East Anglia had a profound effect on the design and construction of agricultural vehicles and implements generally. Not only did local wheelwrights build two-wheeled carts of Scottish design, but the techniques employed in building carts were adopted for the larger four-wheeled wagons. Indeed, even the spectacle design on the frontboard of Scotch carts was incorporated in these wagons. It may be said that the Eastern English plank-sided wagon is, in effect, a four-wheeled version of the Scotch cart, since the wagons of the East Midlands and East Anglia are more closely related to the Scotch cart than to the traditional panel or spindle-sided vehicles of the area. In Hereford, on the other hand, where the Scotch cart did not penetrate until a much later date, its influence is seen only in the planking of the sides of the wagons, for the craftsmen in that county clung to the techniques, sizes and designs employed in the older panel-sided vehicles.

One other feature transferred from the Scotch cart to the plank-sided wagons in Eastern England was the material used in the construction of the body sides. While the older type of vehicle was boarded with ash, poplar or elm, the newer vehicles were always boarded with deal planks.

## NOTES

1. CHILDE, V. G.  'True or Continuous Rotary Motion', *A History of Technology*–Edited by C. Singer, E. J. Holmyard and A. R. Hall–Oxford 1954–Vol. 1, p. 211
2. *Ibid.*–p. 208.
3. *Ibid.*–p. 107
4. *Ibid.*–p. 208
5. *Ibid.*–p. 212
6. EVANS, E. E.  *Irish Heritage*–Dundalk 1941–p. 111
7. MOTEFINDT, H.  'Die Entstelhung des wagens und des Wagen rades', *Mannus*–Vol. X, 1918–pp. 32-63
8. CHILDE, V. G.  *op. cit.*–p. 213
9. *Ibid.*–p. 212
10. BIOT, E.  *La Tcheou Li*–Imprimerie Nationale Paris 1851–Vol. II, Chapter XL, p. 466 *et seq.* Translated from the French by Jean Pace and R. A. Salaman, 1955 (unpublished). The French was a translation of *Khao Kung Chi* (Artificers Record) contributed in the second century B.C. by the Han Prince Te of Ho-Chien after the loss of the original.
11. CHILDE, V. G.  *op. cit.*–p. 212
12. BULLEID, A. and H. ST. GEORGE GRAY.  *The Glastonbury Lake Village*–Glastonbury, 1911–Vol. I, p. 337
13. CURLE, J.  *A Roman Frontier Post and its People*–Glasgow, 1910–p. 292
14. *Ibid.*–p. 244

15. FORBES, R. J. 'Roads and Land Travel', *A History of Technology*–Edited by C. Singer, E. J. Holmyard, A. R. Hall and T. I. Williams–Oxford, 1956–Vol. II, p. 533
16. JOPE, E. M. 'Vehicles and Harness', *ibid.*–p. 548
17. GWEI-DJEN, LU, R. A. SALAMAN and J. NEEDHAM. *The Wheelwright's Art in Ancient China*–MS dated 1957–paragraph 6–'Looking at the cake-like convexity (*ping*) one requires that the *chua* (the tongue of the spoke which enters the felloe) is correctly placed (i.e. meets it with exactitude at right angles).'
     *Ibid.*–paragraph 22–'On a wheel of 6′ 6″, the cake-like convexity (*ping*) amounts to two thirds of an inch . . . When a wheel is dish shaped (*pi*) i.e. domed towards the body of the cart, the vehicle will move without jolting.'
18. JOPE, E. M. *op. cit.*–p. 548
19. GWEI-DJEN, LU, R. A. SALAMAN and J. NEEDHAM. *op. cit.*
20. SMALL, J. *A Treatise on Ploughs and Wheel Carriages*–London 1784–p. 200
21. STURT, G. *The Wheelwright's Shop*–Cambridge 1923–pp. 91-3
22. RICKMAN, JOHN (Editor). *Life of Thomas Telford*–London 1838–pp. xvii-xviii
23. STRATTON, E. M. *The World on Wheels*–London 1878–p. 35
24. [Dr. D. H. Fender has supplied the following note. 'The strength of a simple spoked wheel depends on the resistance offered by the spokes to bending along its length, to twisting and to compression or elongation. Only the first two effects have to be considered for a flat wheel. In the case of a dished wheel, all three are present. A simplified calculation for an average wheel shows that the third effect is some twenty times larger than the sum of the other two. Hence, one might expect the strength of a dished wheel to be increased by this proportion.']
25. [The following table gives the dimensions of the wheels of four wagons—a carrier's wagon of 1780, a farm wagon of 1838 from Oxfordshire, a farm wagon of 1870 from Dorset, and a boat wagon of 1900 from Wiltshire. From the table it may be inferred that the design and size of wheels of the traditional farm wagons were generally based on those of the earlier road wagons. The carrier's wagon and the Oxfordshire farm wagon have wooden axles, and large hubs were therefore necessary to accommodate the thick wooden arms of the axles. The Dorset farm wagon and the Wiltshire boat wagon, on the other hand, have iron axles and consequently the hubs are smaller in size, while the spoke mortices are staggered.

|  | Road Wagon *c.* 1780 | Oxfordshire Farm Wagon 1838 | Dorset Farm Wagon 1870 | Wiltshire Boat Wagon *c.* 1900 |
|---|---|---|---|---|
| **REAR WHEELS** | | | | |
| Diameter . . | 62½″ | 61″ | 53″ | 50″ |
| Number of Spokes . | 14 | 12 | 12 (Staggered) | 14 (Staggered) |
| Width of Tyre . | 4″ | 2½″ | 2½″ | 2″ |
| Diameter of Hub (just in front of spoke mortices) . | 14″ | 14″ | 12″ | 10″ |
| Length of Hub . | 16″ | 15″ | 12½″ | 12″ |
| Method of Tyring . | Hoop | Strakes (6) | Hoop | Hoop |

|  | Road Wagon | Oxfordshire Farm Wagon | Dorset Farm Wagon | Wiltshire Boat Wagon |
|---|---|---|---|---|
| FRONT WHEELS |  |  |  |  |
| Diameter | 45½″ | 52″ | 38″ | 34″ |
| Number of Spokes | 12 | 10 | 10 (Staggered) | 12 (Staggered) |
| Width of Tyre | 4″ | 2½″ | 2½″ | 2″ |
| Diameter of Hub (just in front of spoke mortices) | 13″ | 11″ | 12″ | 9″ |
| Length of Hub | 14½″ | 15″ | 12½″ | 12″ |
| Method of Tyring | Hoop | Strakes (5) | Hoop | Hoop |

26. CHILDE, V. G.   op. cit.–A History of Technology–Oxford 1954–Vol. I, p. 212
27. SINCLAIR, J.   Agricultural Report of Scotland–Edinburgh 1814–Vol. I, p. 717
    HENDERSON, J.   General View of the Agriculture . . . of Sutherland–Edinburgh 1815–p. 59
28. HADDON, A. C.   The Study of Man–London 1898–p. 189
29. CLARK, J. G. D.   Prehistoric Europe–London 1952–p. 305, Fig. 169
30. Ibid.–p. 305, Fig. 168
31. CZEKANOWSKI, J.   'Z Dziejow Wozu i Zaprzegu', Lud (Organ of the Polish Ethnographical Society'–Cracow-Poznan 1952–p. 114
32. PEATE, I. C.   'Some Aspects of Agricultural Transport in Wales', Archaeologia Cambrensis–1935–p. 219 et seq.
33. WORGAN, G. B.   General View of the Agriculture . . . of Cornwall–London 1815–p. 37
    MARSHALL, W.   Rural Economy of the West of England–London 1797–p. 86
34. MARSHALL, W.   Rural Economy of Yorkshire–London 1788–Vol. II, p. 254
35. BERG, GOSTA.   Sledges and Wheeled Vehicles (Nordiska Museets Handlingar 4)–Stockholm and Copenhagen 1935–Plate XXVIII, 3
36. Ibid.–Plate XXX, 1
37. Ibid.–Plate XXIX, 2
38. TUKE, J.   General View of the Agriculture . . . of the North Riding of Yorkshire–London 1800–pp. 78-9
39. YOUNG, A.   General View of the Agriculture of Essex–London 1813–Vol. I, p. 161

# DISTRIBUTION
## III

## 1 CONTINENTAL DISTRIBUTION

In many parts of the world both two-wheeled carts and four-wheeled wagons are used for farm transport. Where both types of vehicles occur, carts are used for one type of load and wagons for loads of a different kind. For instance, in parts of the French Massif Central carts are used to carry dung, potatoes and other root crops, while on the same farms wagons only make an appearance during the hay and corn harvests. On the other hand, in Picardy root crops are carried in wagons, while the hay and corn harvest are transported in large carts.

There are many other regions, however, where only the one type of agricultural vehicle is known, and either carts or wagons are used for all farm transport. In Poland, Czechoslovakia and Hungary, for instance, all farm transport is undertaken by long bodied wagons. On the other hand in countries such as Portugal, India and Indonesia, the only wheeled vehicle known is the cart.

Fig. 10 illustrates the distribution of carts and wagons in Europe and shows that the cart is used in two wide belts on the northern and southern periphery of the continent. The large North European horse-drawn cart is found throughout north-western Europe from Iceland to the Urals, while the smaller, ox-drawn southern cart occurs in a wide belt from Portugal to the Philippine Islands. In France both these zones meet and carts representing the fusion of the southern and northern varieties are used throughout the country.

While the cart clings to the peripheries of Europe and Asia, the wagon occurs in a vast mid-continental domain extending from Holland and Flanders in the West, to Kamchatka on the Pacific coast of Russia in the East. Throughout this huge tract of land the wagon has remained the only form of wheeled vehicle. In Europe it is used exclusively in Poland, Austria, Czechoslovakia, Hungary and Denmark. Except for a small region in the Lower Rhine Basin, where carts of the large northern type made an appearance in the nineteenth century[1] the wagon is used throughout Germany.

In Scandinavia the wagon has encroached on the southern

margins of the various countries and ousted the northern cart from those regions. Berg[2] points out that in Sweden, the boundary between a northern cart-using region and a southern wagon-using region passes approximately through the centre of the country, 'separating the more conservative parts of Sweden from those in which innovations had gained a footing, and been accepted before

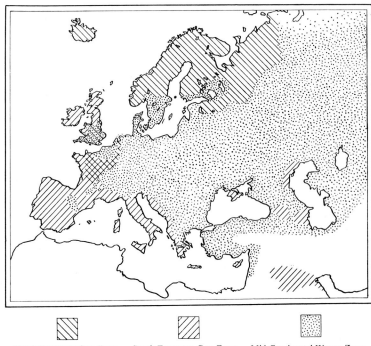

North European Cart Zone     South European Cart Zone     Mid-Continental Wagon Zone

10   The distribution of carts and wagons in Europe

older ideas were abandoned'.[3] In other Scandinavian countries the pattern is repeated, for in recent years the wagon has encroached on the shores of Oslo Fiord and the lower Glommen Valley, while it is used in southern Finland, in Aland, Karelia and parts of the south west. With the exception of parts of Esthonian Ingria, wagons are used throughout the old Baltic states, and they occur in all parts of European Russia except the north-west, and small areas on the southern periphery such as the Crimea and the shores of the Caspian Sea.

In western Europe the wagon is used in all parts of Holland except a narrow strip of country running from the Friesian Islands, through Gelderland, North Utrecht and Limburg, where the northern cart replaced wagons in the eighteenth and nineteenth centuries. From Limburg this narrow belt of cart-using land extends to eastern Belgium, but throughout the remainder of the country wagons are used.

As one might expect in a country which has long been the melting pot of diverse cultural elements, the pattern in France is extremely complicated. Here, not only does the large northern cart meet the smaller southern variety, but the wagon and even a three-wheeled farm vehicle also make an appearance in a most complex pattern of distribution. Although carts are used in the greater part of France there are numerous isolated regions where only wagons are found. The stronghold of the wagon is the 'continental' provinces of Franche Comté, Lorraine and Savoy, where long, narrow-bodied vehicles with low wheels are used for all farm purposes. These wagons with bodies sloping outwards from a narrow floor are similar to the wagons used throughout East and Central Europe.

In addition to this solid block of wagon-using country in eastern France, wagons may still be found in isolated Morvan, Forez, Haute Auvergne and Gevaudun. In Limousin, too, the wagon still persists in the more remote districts, while in Limagne, Bourbonnais, Rouerge and Velay it is used alongside carts for harvesting duties. These regions where wagons persist are remains of a wider area which extended southwards over the Spanish border. In Spain, wagons have been introduced during the last hundred years for harvesting olives and cereals in Aragon, La Mancha and Andalusia.[4] According to the age-old custom of Mediterranean Europe these wagons are harnessed to ox teams.

In France there is evidence that the wagon province has shrunk considerably during the last hundred years, so that what may have been a wagon-using 'promontory' stretching south from Lorraine, has been broken into 'islands' by the advance of the cart.[5] In Limousin, for example, the wagon now occurs only on the isolated plateau of Millevaches, but in the eighteenth century it was found at St. Leonard, and in the seventeenth century occurred on the banks of the Loire in Touraine.[6]

In addition to the long, narrow-bodied typical Central European

wagons of eastern France and the heavier, shorter wagons of the
Massif Central, very large, heavy wagons are found in Flanders,
Picardy, Boulonnais and Brie. In this north-eastern region
peculiar three-wheeled wagons are also used, mainly for carrying
root crops. The chief vehicle of agricultural transport in Latin
Europe is the ubiquitous ox cart, but in northern Italy, in
Lombardy, Piedmont and Emilia, the wagon is also found. It is
used throughout Switzerland, from where only in relatively recent
times it has penetrated the Alpine Passes to bring a cultural feature
of Central Europe to the northern parts of Latin Europe.[7]

Wagons are used throughout the Balkan Peninsula, many of
them in the south being ox-drawn, but in the provinces bordering
the Adriatic Coast, namely Herzegovina, Dalmatia, Epirus and

11   (a) Modern European wagon, U.S.S.R.

Albania, ox carts are used.[8] From the Balkans, the wagon penetrates
eastwards into Anatolia as far as Aleppo and Hama, where it meets
a region of nomadic pastoral life without wheeled vehicles. East of
the Black Sea the wagon zone extends southwards to Tabriz in
Persia and eastwards to Turkestan and even to the shores of the
Bering Sea. In Asia its southern boundary is marked by a line
passing approximately through northern Sinkiang, Mongolia and
Manchuria.[9]

11 (b) Modern European wagon, France

## 2 BRITISH DISTRIBUTION

In the British Isles there are no regions where the wagon alone is used for all farm transport, although there are certain districts, such as Dorset and South Shropshire, where two-wheeled carts are uncommon. With a few exceptions, carts are used in all parts of Britain; in the Celtic north and west they are used for all farm transport, while on the English Plain they are limited to carrying loads of dung, loose gravel, sand and similar material. Wagons, however, are limited to the south-eastern quadrant of the British Isles, where they are widely used for carrying light but bulky loads such as hay and corn sheaves. The distribution of the wagon in the British Isles (Fig. 12) suggests a typical south-eastern cultural pattern. In common with other features originating deep in the continent of Europe[10] its use has spread from the south-east to all parts of the so-called Lowland Zone of Britain.

In western or Celtic Britain wagons are unknown except for those areas in south-western England and Wales which, by their very position, must be regarded as westward extensions of the English Plain. Wagons are sometimes found on the larger farms of Cornwall and Devon; these are closely related not only to the bow wagons of Wessex and the South Midlands, but also to the older two-wheeled Cornish wains. Indeed, so small and light are these Cornish and Devonian wagons that they may be regarded as four-wheeled versions of their ox-drawn predecessors. Even so, wagons are not frequent in the south-west, and those that do occur are found only on large farms in the more level districts.

In South Wales the Vale of Glamorgan represents an extension of English Lowland culture into Wales. Throughout the centuries the Vale has been more closely related to the West of England than to the mountainous centre of Wales. Here Roman settled life was highly developed, while the rest of the country was kept in order by military roads and forts. Here, too, the Anglo-Saxon open field system and associated village life was found, while the remainder of Wales still clung to its run-rig system and tribal, non-village organisation. The Vale today with its trim villages, beautifully thatched cottages, elaborate churches and an Anglicised social climate is far removed culturally from the mountainous heart of Wales. As an element in this westward extension of English Lowland life one finds the bow wagon, a beautifully designed vehicle closely related to the wagons of Gloucestershire, and indeed wagons of a degenerate style are found even in South Pembroke-shire, another area whose culture differs from that of other parts of Wales.

The broad eastern face of the Welsh mountains overlooks the wide curve of the Severn Valley and is entered by several wide vales. Along these natural routeways, the valleys of the Severn, Wye, Usk, Monnow and others, countless invaders have made their way, and with them countless ideas from the English Plain have passed through, penetrating into the very heart of the highlands. An instance of this is the appearance of the characteristic half-timbered houses of Shropshire along the Severn Valley. These houses, far removed from the plain unornamented architecture of Welsh rural homes, are found as far west as the shores of Cardigan Bay. In the same way the practices and tools of agriculture of these eastward facing tongues of lowland, are far more closely related to those of the English Plain than of the Welsh Highlands. As an element in this cultural influence, wagons based on Shropshire design occur as far west as the Dovey estuary.

Wagon types in the whole of Wales, in the south, centre and north, must be regarded as intrusions from the English Plain.

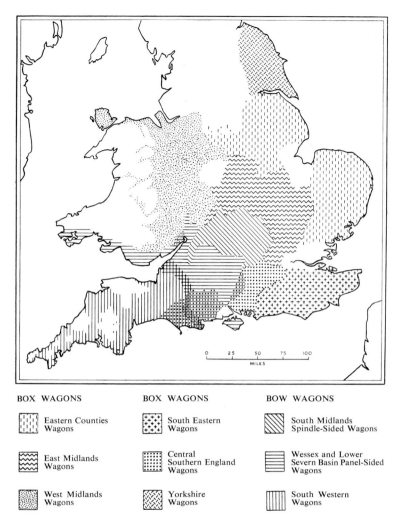

BOX WAGONS

- ⊞ Eastern Counties Wagons
- ⌇ East Midlands Wagons
- ▦ West Midlands Wagons

BOX WAGONS

- ⊞ South Eastern Wagons
- ⊞ Central Southern England Wagons
- ⊠ Yorkshire Wagons

BOW WAGONS

- ⧄ South Midlands Spindle-Sided Wagons
- ☰ Wessex and Lower Severn Basin Panel-Sided Wagons
- ⦀ South Western Wagons

12  The distribution and regional types of English farm wagon

They have intruded from the West of England into the Monmouth-shire coastal plain, the Vale of Glamorgan, and as far west as the Atlantic coast in South Pembrokeshire; from Hereford and Shropshire into the valleys of the eastward flowing streams, and from Cheshire along the North Wales coastal plain as far west as Anglesey.

In Scotland wagons were never built. Sinclair, writing in 1812[11]

said that at that time there were no more than six wagons in the whole of the country and even these had been introduced by English landowners. Although wagons were introduced into Ireland by prosperous land owners during the period of agricultural improvement in the eighteenth century, those vehicles were soon abandoned in favour of the traditional low-back and trottle cars.[12] In the Lake District, Lancashire, Northumberland and Durham, as well as in the greater part of Yorkshire, no wagons were known until mass-produced trolleys appeared in some parts, notably in South Lancashire and the West Riding of Yorkshire in the last decade of the nineteenth century. The only districts in the North of England where wagons were known before that date were the East Riding of Yorkshire and some of the Dales of Derbyshire where wagons of Lincolnshire type were found on large farms in the more level districts.

Thus from the point of view of distribution the wagon has a definite and important place in the agricultural life of the English Plain. However it is entirely unsuited to the economy and topography of Highland Britain and is not found there.

## 3    REASONS FOR WAGON USE

In the foregoing sections the geographical distribution of agricultural vehicles in Europe and Asia was described. In Fig. 10 it may be seen that carts are found in wide belts on the periphery of the Eurasian continent while wagons are limited to a broad mid-continental sector extending from the shores of the Bering Sea to the North Sea.

Why is there this peculiar distribution pattern? Why, as far as Britain is concerned, should the wagon be such an important feature in the agricultural life of the English Plain while it is almost completely absent from the farms of Highland Britain? There are five factors, of greater or lesser importance, which may explain this distribution and each factor will be examined in turn and its validity considered. These are: (1) Topographical factors, (2) Economic factors, (3) The size of holdings, (4) Field pattern, (5) Cultural factors.

### (1)    TOPOGRAPHICAL FACTORS

The distribution of carts and wagons, in Britain at least, does suggest that topography plays a very important part in determining the type of vehicle used in different areas.

In Highland Britain there are large and continuous stretches of land more than 1,000 feet above sea level, with numerous mountains, deep narrow valleys and steep slopes. Stretches of level land are strictly limited in extent, although most of the settled life of the zone is concentrated in these intermont and coastal plains. The broken character of relief, the steep slopes and poor soils all make farming difficult in Highland Britain. This topographical character prevents the use of large and heavy agricultural vehicles and in some parts prevents the use of wheeled vehicles altogether. In many regions, such as Central Wales and North Scotland, wheel-less sledges are still used since even small carts are unmanageable on the very steep slopes of these regions. In other parts of Highland Britain such as Cornwall and Cumberland, pack animals were used for local transport until recent times.

Lowland Britain on the other hand provides a striking contrast to the Highland Zone. With the hills seldom rising above a few hundred feet, there are large stretches of level land and settlements are dotted evenly over the surface of the country. The environment is much kinder to man, the soils are deeper, and there are few steep slopes to interrupt cultivation. Since Lowland Britain is relatively level, large agricultural vehicles can be used almost everywhere with little danger of collapsing on steep slopes; and topography is not the controlling force that it is in Highland Britain.

The British evidence suggests that topography is an all-important factor in determining the type of vehicle used in the various regions. It may be argued that large unwieldy wagons would soon collapse on the steep slopes of Highland Britain, and for this reason small carts, sledges and pack animals have predominated in the zone. A closer examination, however, shows that the correlation between wagons and level land is by no means exact. For example, in South Shropshire one would expect light two-wheeled carts in a region which is topographically and geologically related to the mountains of Wales. In fact, in this hilly area the wagons are extremely heavy while carts are almost unknown. On the other hand, North Shropshire, which is much flatter than the south, has very few wagons and farm transport is undertaken almost entirely by carts. The Dovey Valley in Montgomeryshire, cannot by any stretch of the imagination be called flat, and yet wagons of Shropshire type are found there.

In Europe the influence of topography on vehicle types is even less pronounced. Wagons are found throughout the Carpathians, the Alps and the Jura. They occur in hilly Franche Comté and the rugged parts of the Massif Central, but in the more level districts of Normandy and Aquitaine the cart is widely used and wagons non-existent.[13]

It seems, therefore, on closer examination that topography is only a minor factor in determining the type of vehicle used in a particular region. Although topography may decide initially whether wheeled vehicles are practical at all, it does not determine whether the vehicles used should be two-wheeled or four-wheeled.

It does seem, however, that topography sometimes influences the design and construction of wagons in a particular region. For example, in Dorset, a county full of abrupt changes of slope, wagons are small and light, while in the hilly Cotswolds they are shallow bodied with excess weight cut away wherever possible. East Anglia shows the influence of topography in reversed circumstances where the land is level and the vehicles large and heavy.

## (2)  ECONOMIC FACTORS

The main hill masses of Britain occur in the north and west of the country. The hills of Highland Britain stand in the path of the rain-bearing winds which blow from the west, hence the rainfall there is heavy. Lowland Britain, being situated to the east of these hill masses is comparatively dry.

Heavy rainfall combined with the general poverty of the soil in Highland Britain make the zone quite unsuitable for cereal cultivation, and throughout the centuries pastoral farming has been the basis of the economy. Large scale cereal cultivation is concentrated in the fertile, warm and sunny counties of Lowland England.

In Britain it has been the custom to use four-wheeled farm vehicles for harvesting corn, and so wagons are numerous wherever there is large scale crop raising. In East Anglia, for example, cereal cultivation is the basis of the economy, and wagons are very common on the farms of Norfolk, Suffolk and North Essex. On the other hand, Wales has very few wagons since the economy is overwhelmingly pastoral and crop raising is of minor importance.

Furthermore, economic factors play an important part in

determining the character of farm transport within Lowland England itself. Where the economy is based on cereal cultivation, wagons are plentiful, but in those regions where the economy is based on pastoral farming, wagons are rare. On the Cheshire Plain, North Shropshire and parts of Somerset, where grassland predominates, wagons are by no means common. In those regions carts, mainly of the Scotch variety, are used for all farm purposes including carrying the hay and the limited amount of corn grown on each farm. Whereas in East Anglia wagons are plentiful, on the corn growing boulder clay of Norfolk, Suffolk and North Essex, they become almost non-existent in the market gardening and dairying region on the clay of South Essex.

Although as a rule wagons are used primarily for harvesting hay and corn, this is not invariably the case. In the counties of Radnor and Montgomery, the harvest is carted by two-wheeled *gambos*, while wagons are used exclusively for carrying root crops.

In Europe the association between wagons and hay and corn harvesting is by no means as close as it is in England. In Central and East Europe, of course, wagons are used for all farm purposes, but in districts which possess both wagons and carts, custom has dictated that each type of vehicle should be employed for a specific kind of load. Over much of France long narrow wagons fitted with end ladders are used for harvesting corn and hay, as in England, while tip carts are used for root crops and manure. In Picardy, however, wagons are used exclusively for carrying root crops and manure while two-wheeled carts are limited to harvesting hay and corn.

## (3) SIZE OF HOLDINGS

The distribution of wagons in the British Isles suggests that the extent of their use is influenced by the size of holdings. Wagons are limited to the south-eastern quadrant of Britain, to an area of large farms where the arable fields may be some distance from the farmstead. On the other hand Celtic Britain, where smallholdings predominate, has no wagons.

In Celtic Britain the farms are small and all the fields are within easy reach of the homestead, so that there is no demand for numerous large vehicles. Sledges and small carts are sufficient for the limited transport needs of the upland farmer. Moreover, where there are a large number of smallholdings on inferior land, the income of the individual farmer is relatively low and he cannot afford elaborate tools and implements. In such a society specialised craftsmen are rare, and from time immemorial the upland farmer has made his own tools and implements from materials found on his farm. This is one reason for the continued use of primitive sledges and hand barrows in Highland Britain.

While the smallholding predominates in Highland Britain, in Lowland Britain the farms are much larger and the arable fields may lie a considerable distance from the farmstead. For example, in East Anglia, farms of 1500, 2000 and even 2500 acres are by no means rare. A typical farm on the Wiltshire Downs has between 500 and 1000 acres. Wagons are plentiful in these two regions, as they are in other areas where farms are large. This is very clearly illustrated in a relatively small area in the case of Hampshire. The north of the county with its large cereal growing farms has numerous wagons, while in the south where farms are small, wagons are extremely rare. Again, in Wales, wagons are found in those areas where large holdings occur. Radnor, with 61 per cent. of its farms over 100 acres, and Glamorgan, with 45 per cent. over 100 acres, are both counties where the wagon is common. This is in sharp contrast to the upland core of Wales where smallholdings and the use of primitive agricultural vehicles have persisted.

## (4) FIELD PATTERNS

Over the greater part of the English Plain, including those outposts of Lowland culture in Highland Britain, some form of open field system was prevalent. This is also the area where the wagon is most used. Is this distribution due to the difficulty experienced in using wagons on enclosed fields in the Highlands and the relative ease with which even vehicles with a limited lock could be manoeuvred in the open fields of the Lowland? It is unlikely, and the probable explanation is that the open field system and the wagon were both called into use by the same economy, that of the level, crop-raising English Plain facing the Continent, and are not dependent on each other.

On the European continent, both the wagon and open field system of agriculture occur in a wide mid-continental belt which runs from the shores of the North Sea as far as Poland and Russia, including the southern parts of Sweden. The wagon seems to occur over a much wider area than the former open fields, for in addition to the wide belt of country mentioned, it penetrates into the Alpine Provinces of Europe, to the Balkan Peninsula and North Italy, though dispersed settlements characterise these areas. Although the initial spread of the wagon in the North European Plain may have been to areas of open fields and nucleated settlement, it has spread subsequently outside the bounds of this economy. Hence the correlation between the two elements, if it ever really existed, is no longer valid.

## (5) CULTURAL FACTORS

West and North Britain derive many cultural features from the Mediterranean region of Europe, and agricultural vehicles of Mediterranean type occur everywhere within Highland Britain.[14] South-East Britain, on the other hand, derives many of its cultural features from the middle sector of the European continent. 'South-East Britain forms the apex of a triangle pivoted on Cap Gris Nez, presenting two sides to a long strip of continental coast from Western France to the Danish peninsula . . . cultures arriving . . . tended to spread generally over the Lowland Zone giving to their associated objects a characteristic south-eastern pattern.'[15] The wagon originated in Central Europe and spread gradually to the whole of the North European Plain, replacing the cart on the northern fringes of Mediterranean Europe and on the southern fringes of Scandinavia. From the Low Countries it was introduced into Britain, and despite its late arrival in our island it spread rapidly throughout the Lowland Zone to give a typical south-eastern cultural pattern.

It must be remembered, however, that Lowland Britain, unlike most of the North European Plain, is not exclusively a wagon using area, since two-wheeled carts are also found there. This raises the question of how these carts fit into the pattern of distribution. The Lowland cart differs considerably from its western and northern British counterpart; while the Celtic carts are light, those of the English Plain are heavy and box-like. One horse or a pair of oxen is enough to draw the wains and gambos of Celtic Britain, but the tumbrils and box carts of the English Plain need at least two horses. There are also many differences of construction.[16] All this suggests that one must look to some place other than Mediterranean Europe for the origin of the English tumbril.

Throughout the middle and northern provinces of West Europe

where carts are used, they are all similar in construction to the tumbril. In Scandinavia, in parts of the Netherlands, the Lower Rhine Valley in Germany, Normandy, the Paris Basin and the northern parts of the Massif Central, large carts are widely used. It is probable that in the past the box cart was used over a much wider area than at present. It is only within living memory that the wagon replaced the cart in Southern Norway and in Picardy wagons have become common only during the last hundred years. The occurrence of box carts in the middle sector of the European continent in the past would suggest that as far as Britain is concerned, too, the distribution of the northern cart should display a south-eastern cultural pattern. This is in fact the case, for in design the tumbril is far removed from the carts of Mediterranean origin used in Highland Britain.

The wagon, which is used alongside the tumbril in south-east Britain, was introduced from the continent in post-medieval times. In detailed design and construction, however, British wagons are vastly different from those used in the North European Plain. While the so-called eastern or continental wagon has four small wheels of near equal size, those of Britain have large rear wheels and small fore-wheels. Many eastern continental wagons have braces running from the naves to the top-rail of the body, but no such construction is known in Britain. The most common type of body on continental wagons consists of a number of long spindles projecting outwards from the main frame. The sides are unboarded and the long, narrow body is almost V-shaped. The bodies of British wagons, on the other hand, have vertical sides and are boarded, and much wider, while the majority are equipped with sideboards to take the overhanging load. It is only on the western periphery of Europe, in Scandinavia and the Low Countries, that wagons remotely resembling their British counterparts are found. In Holland, for example, while the majority of wagons are of the eastern open-sided variety, a few are found with wider, boarded-in sides, and having rear wheels larger than the fore-wheels. These few vehicles represent the transition between the open-sided, eastern wagon and the true English wagon. Though these vehicles have the boarded bodies of the English type, the complete absence of sideboards and the long narrow bodies show their close relationship to their eastern ancestors.

Once again the ingrained characteristic of Lowland Britain is reflected in the character of its agricultural transport. 'The ultimate expression of any continental culture in Lowland Britain tends to possess individual characters', says Fox.[17] 'The sea barrier inhibits mass movement and encourages independent adventure; Lowland culture at any given period thus tends to represent the mingling of diverse continental elements, rather than the extension beyond the straits of a single continental culture.' The English four-wheeled farm wagon is an example of this *par excellence*.

## NOTES

1. [Dr. J. M. G. van der Poel of Landbouwhogeschool, Wageningen, has drawn my attention to the fact that carts were not widely used in the Lower Rhine Basin until at least the mid-eighteenth century. Since 1850 the cart has rapidly replaced wagons in the region between the Meuse and the Waal.]
2. BERG, GOSTA. *Sledges and Wheeled Vehicles* (*Nordiska Museets Handlingar 4*)–Stockholm and Copenhagen 1935–p. 110
3. *Ibid.*–p. 110
4. Snr. Vieves de Hoyos of the Museo del Pueblo, Madrid, private communication
5. DEFFONTAINES, P. 'Sur la repartition geographiques des voitures a deux roues et a quatre roues', *Congress International de Folklore*–Paris 1937– Translation by D. Paton (unpublished)–p. 5
6. *Ibid.*–p. 4
7. *Ibid.*–p. 8
8. Professor Branimir Bratanic of Zagreb, private communication.
9. [It is unnecessary to consider the distribution of wagons outside the continents of Europe and Asia. In other continents the vehicles were introduced fairly recently by settlers and followed the style of construction used in their country of origin. For example, the well known Boer wagon of South Africa is similar to that used in the Low Countries, while the equally famous Conestoga wagon of North America was based on the large vehicles of Western Germany and the road wagons of eighteenth-century Britain.]
10. BOWEN, E. G. 'Prehistoric South Britain', *An Historical Geography of England before* 1800 *A.D.*–Edited by H. C. Darby–Cambridge 1951– pp. 7–8–There are many examples of this pattern of cultural distribution in the historical and archaeological record. A prehistoric example is provided by the distribution of Beaker folk burials in the Early Bronze Age. The Beaker folk, who probably originated in the Rhine Valley, came to Britain *via* the south-east coast, the Wash and the inlets and estuaries of the north east. From these districts of invasion they spread inland to cover a great deal of Lowland Britain. Like all other Lowland distribution, Beaker burials are found in those districts which may be regarded as appendages of the English Plain—the North Wales seaplain, the Vale of Glamorgan and the Severn Valley.
   WOOLDRIDGE, S. W. 'The Anglo-Saxon Settlements', *Ibid.*–p. 88 *et seq.*–A later example is provided by the distribution of Anglo-Saxon settlements in Britain. The Anglo-Saxons, who originated in the middle sector of the European continent, came to Britain *via* the eastern coast and spread their influence over the whole of the English Plain. The standard type of economy that they introduced, a village economy with the tillage of the soil and the rotation of crops in open fields, became

E

prevalent in the greater part of the Lowland Zone. The open field system penetrated along the traditional routes of invasion into the heart of Wales and Cornwall.

11. SINCLAIR, J. *Agricultural Systems of Scotland*–Edinburgh 1812–p. 71
12. EVANS, E. E. *op. cit.*–p. 111
13. DEFONTAINES, P. *op. cit.*–p. 1–'The four wheeled wagon is longer and narrower than the two wheeled cart. The weight being distributed over four points allows lighter wheels . . . the two wheeled cart can turn in one spot, but it raises the problem of balance; to have stability it must be wider and shorter. As the weight rests only on two points, the cart requires stronger wheels.' [In France the wagon is better balanced and light in construction and is therefore quite suitable for use in hilly regions. In Britain, on the other hand, where the wagon is much heavier and the wheels are large in diameter, it is less stable than the cart in hilly country.]
14. JENKINS, J. G. 'Two Wheeled Carts'–*Gwerin*–Vol. II, No. 4, 1959– pp. 162-75
15. BOWEN, E. G. *op. cit.*–p. 7
16. JENKINS, *op. cit.*
17. FOX, C. *The Personality of Britain*–Cardiff 1932–p. 88

PART TWO

# STRUCTURE

# BUILDING

## IV

### 1  WHEELS

The craft of wheelwrighting is one of the most complicated of all woodcrafts. In many respects it is similar to hardwood joinery, but whereas the joiner makes his joints to fit, relying a great deal on glue, the wheelwright relies on tightness of joints alone to hold his work together. He uses many of the same tools as the carpenter with a few others peculiar to his craft. In the past the wheelwright was an essential member of all village communities but today, with the disappearance of horse transport, very few still remain. Those that still practice the craft tend to spread their activities over a much wider field; hence one finds craftsmen who in addition to being wheelwrights are also joiners, carpenters and undertakers.

A wheel consists of a central nave, stock or hub, from which an even number of spokes radiate. The felloes are the curved members forming the rim of the wheel. An iron tyre, or a series of crescent-shaped pieces of iron, called strakes, bind the whole wheel together.

### (1)  NAVE

The nave[1] of a wheel is nearly always made of elm, although less satisfactory ones have been made of oak or ash. Elm has a number of advantages over other timber, the most important being the nature of its twisted grain which resists any attempt at splitting or cleaving. The timber must be able to withstand very heavy hammering when the spokes are driven into the nave, and elm is one of the few timbers that does not split as a result of this treatment. Elm trees are rarely localised in natural forests or plantations, but are well distributed throughout Britain and wheelwrights in all parts of the country have a plentiful supply of timber close at hand.[2]

First of all the wheelwright buys straight, smooth, winter felled elm trees and transports them to his yard. With a cross-cut saw the timber is sawn into lengths of 14 or 15 inches. A hole is bored right through the centre of each block of wood with an auger in order to assist the drying process. The elm butts are then stored

HAPPY THE MAN WHOSE WISH AND CARE CONTENT TO BREATHE HIS NATIVE AIR
A FEW PATERNAL ACRES BOUND, IN HIS OWN GROUND. POPE

13  The wheelwright, an engraving by Stanley Anderson, CBE, RA

with the bark still on until they are thoroughly dry. Every three
or four months the wheelwright visits the store to inspect the butts,
and the mildew that appears on the elm as the sap is slowly exuded
is brushed off with a hard brush for if this is not done the wood
will gradually rot. For a hub 12 inches in diameter at least six
years are required for seasoning, while for the larger naves, 15 inches
in diameter, the seasoning period may be as long as ten years.

In some parts of the country it was customary first to place the roughly sawn elm butts in a running stream so that water replaced the natural sap in the wood, after which the drying process was hastened considerably.

When the seasoning is complete and the bark has been chopped off with an axe, the perfectly dry naves are placed in a lathe and turned to the required shape and diameter Usually this hand-powered lathe is of special design.[3] A driving wheel 6 feet or more in diameter rests firmly in a framework some distance from the lathe bed and at right angles to it. The bed may be anything up to 12 feet in length and firmly fixed to the ground by a set of thick wooden legs at each extremity. Along the centre of the bed is a crack in which run the puppets for pivoting the nave. The puppets, which may be adjusted anywhere according to the length of the material being turned, are then screwed tightly to the correct length by a pair of bolts on the underside of the bed. The driving wheel is connected to the pulley wheel of the lathe mandrel by a strap, and while one craftsman turns the wheel by means of the iron crank, the other, with gouge and chisel, works on the revolving wood, resting the tools on a T-shaped support. In turning a nave only two types of lathe tool are required—a gouge, some $1\frac{1}{2}$ inches in width for the first rough turning, and a chisel of approximately the same width to smooth and cut the line where the iron hub-bands will later be fitted.

As the butt revolves the wheelwright continuously checks it with his calipers to ensure that too much is not being taken off. These calipers, with semicircular, in-curved arms, are generally large and are indispensable when turning.

The nave can be turned in one of two patterns, either 'coach pattern' to accommodate three iron bonds, or 'wagon pattern' with only two bonds. On the wagon pattern nave, the hind bond is $1\frac{1}{2}$ inches wide and $\frac{3}{8}$ inch deep, and is fitted flush with the end of the nave. The front or breast bond, on the other hand, is fitted on so that its inside edge is $\frac{3}{4}$ inch in front of the spokes. After turning, the nave should have two gauge marks on it. The first of these, 8 inches from the hind end, acts as a guide for the front end of the spoke mortices; while the other marking line, $\frac{1}{2}$ inch behind the 8 inch line, is to allow for the staggering of the spokes The spokes of the later type of wheel are staggered so that one line of spokes have their front ends in line while each alternate spoke has its front end a little behind its neighbour. If the spokes were fitted in one line, then the mortices would weaken the nave—a very severe defect when using the modern naves of small diameter. In addition, the back line of spokes has the effect of bracing the wheel.

After turning, the next task on the naves is to saw off the ends to the correct length and to trim them up. Particular care is required with the front end, since the whole alignment of the wheel is regulated from this end.

The nave is now placed on a low stool called a morticing cradle, and is held there by two pairs of wedges, a pair on either side of the nave. A pair of compasses is taken and the arms set at roughly the required distances apart in order to prick off the centre of the spoke mortices on the gauge line. The arms are adjusted by trial and error until all the mortices are marked. When this is done, the prick marks are numbered to ensure that all the mortices are accounted for. A small tee-square is taken and placed at the prick marks square on the gauge line. A pencil mark is drawn back from the line to get the centre of the spoke mortice.

A spoke is now required to mark out the mortices on the nave. By holding the 3 inch by 1 inch foot of the spoke against the gauge line, and exactly over the pencil line, each mortice is marked and its outline accurately drawn with a pencil. This is the exact size of mortice to be made.

The mortices are then bored, the wheelwright using a 12 inch or 14 inch sweep brace with a 1 inch bit. Three holes are bored, one at the front, one at the back and one in the centre of each mortice mark. The wheelwright depends entirely on his own judgement and skill during this boring operation, for the front hole has to be at a slight angle to allow for the dish of the wheel, while the back hole should be bored with a little more bevel away from the face of the wheel ensuring that the bottom of the mortice is no more than $2\frac{3}{4}$ inches long. The centre hole is bored to make the removal of the core easier.

When the naves have been bored, the next step is to chisel out the spoke mortices, but before this is done, the spoke set gauge is adjusted. This consists of a piece of hardwood 2 feet 9 inches in length, $2\frac{1}{2}$ inches wide and 1 inch deep with a hole 4 inches from one end large enough to take a $\frac{7}{16}$ inch coach screw comfortably. At the other end a series of $\frac{3}{8}$ inch holes, each 1 inch from the next are cut, and a piece of whalebone 9 inches in length is passed through the required ⅜ inch hole and firmly wedged there. The exact position of the whalebone depends on the actual size of wheel being made. For example, to make a wheel 4 feet 10 inches in diameter the whalebone should be 2 feet 1½ inches from the coach screw pivot; that is, the position of the whalebone should be at half the diameter of the wheel (2 feet 5 inches) minus the depth of the felloes (3½ inches). The whalebone must, of course, be

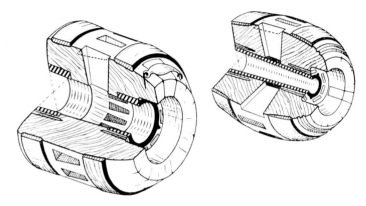

14   Barrel-shaped wheel nave (left) and modern cylindrical type (right)

adjusted to the shoulder of the spoke, rather than to its top. With the spoke set gauge prepared the next step is to plug up the central auger hole that runs through the hub. With a pair of compasses the exact centre of the front end of the nave is found and, at that point, the plug is bored to a sufficient depth to take the coach screw pivot comfortably. The coach screw is screwed up tightly to the face of the nave so that it is just possible to turn the stick without turning the hub. The main purpose of the spoke set gauge is to measure the dish of the wheel, and since the spokes in a dished wheel emerge from the nave at an angle, slanting outwards from the centre, it follows that the spoke mortices have to be cut at a slight angle.

The next step is to measure the horizontal distance between the spoke set gauge and the turner's gauge line on the nave. In a nave of 12½ inches long, this will be 4½ inches, as the face mark is 8 inches from the inside edge of the nave. If one required a ½ inch dish on the wheel, then the whalebone projects 4 inches beyond the stick, and is wedged there securely. In cutting the mortice, the slant of the boring must be such that a small straight edge held against the front of the mortice just touches the whalebone.

A number of tools are required to prepare the mortices. The first of these is a morticing chisel or bruzz—a long socket-handled, V-shaped tool with a blade $\frac{5}{8}$ inch wide. This is used for cleaning the corners of the mortice. Next, a 2 inch firmer chisel is required to cut the core of the mortice, while a $\frac{3}{4}$ inch and a 1 inch heading chisel are used for the delicate shaping of the back and front ends of the mortices. While the bruzz and the firmer or forming chisels are used by pounding with a mallet (which has a head at least 6 inches long, 5 inches wide and 3 inches deep), the more delicate heading chisels are never struck, but worked by hand pressure. Great care is required when using the heading chisel, for even the smallest paring in the mortice makes a great deal of difference at the whalebone gauge. The gauge must be kept in position during the whole operation. In addition to all these tools the wheelwright requires a small inside caliper which is used constantly to check on the size of mortice as the work proceeds.

When all the mortices have been cut the nave is taken to the blacksmith to have the bonds shrunk on. With the older type of barrel-shaped nave, however, it was not customary to fix the bonds until the whole wheel was finished. A bond is put on while still white hot and driven down into place. Water is poured on causing contraction and three small iron pegs are driven into the nave and clenched over the iron bond. Since the old type of nave was barrel-shaped, each bond had to be slightly conical, but the process of fitting bonds to cylindrical hubs is a far less complicated task.

## (2)  SPOKES

No other part of the wheel bears greater pressure than the spokes and for that reason only well seasoned cleft heart of oak is suitable for spoke making. Although some modern spokes have been made of sawn oak they are far from satisfactory for the grain must be unbroken and the cleavage must follow that grain. Cleaving is done by woodland craftsmen, who, after cutting the oak trunks to length with a cross-cut saw, cleave the oak with a beetle and wedge. This must be done during the summer months when the oak is full of sap and the cleavage will run cleanly from end to end.

On arrival at the wheelwright's yard the cleft pieces of oak are roughly dressed with the axe to the required shape and size. They are then stacked in the wood store to season for four years or more. When required, the craftsman takes the spokes and planes them down to a diameter of $3\frac{1}{4}$ or $3\frac{1}{2}$ inches. The foot that fits into the hub is tapered with tenon saw and broad chisel until the tenon measures some 3 inches long and 1 inch wide. With the foot prepared, the exposed part of the spoke is shaped, first with the axe, then with the draw knife and spokeshave, and finally the back is rounded with a hollow bladed plane known as a jarvis. Since the chief strain will be felt at the back of the spoke, great care must be taken not to reduce it too much. The final trimming of the spokes will take place when the whole wheel is assembled.

The nave is next placed over the wheel pit in order to drive in the spokes. This pit is rectangular in shape and measures some 6 feet in length and 10 inches in width. The sides are bricked or lined with timber to make a solid frame for the nave when the spokes are hammered into place. While one person holds the nave steady and keeps the spoke in an upright position the other swings a 14-pound hammer to drive them into place. After each two or three blows the spoke set gauge is pushed into position to make sure that the spoke is being driven in at the right angle. Since the spoke tenons are tapered each blow of the sledge makes the spokes tighter in the hub, and therefore progressively more difficult to correct. If the spoke is being driven in at the wrong angle despite the wheelwright's efforts to correct it by placing his hammer blows carefully, then it must be left until the adjacent spokes have been fitted. In a nave which has two lines of spoke mortices, the front line must be fitted first, and left without final fixing when each spoke is about $1\frac{1}{2}$ inches from its eventual position. Since all the spokes will meet at the centre of the nave, it may be found that the back spokes will be obstructed by the front line from entering the mortices. It is customary then to pass a 2 inch chisel down each side of the mortice and pare away the feet of the obstructing spokes.

To correct the alignment of a spoke, a bridle or crooked stick is used. This, as its name implies, is merely a piece of curved ash 2 feet 6 inches long, and it is placed behind the spoke that is to be driven forwards with its ends in front of the adjacent spokes. In this way the front spoke is pushed forwards to touch the whalebone gauge as it is hammered with the sledge. When all the spokes are in place the wheel is measured with the spoke set gauge to ensure that it has the right dish.

The next step is to set out the tongues of the spokes. When it was customary to tyre wheels with a series of strakes, spokes generally had square tongues; the square-tongued spoke being much stronger but more difficult to shape than the round-tongued variety. A scribe which consists of a piece of ash 2 feet long and $\frac{3}{4}$ inch square is taken, and a bradawl inserted to produce the gauge tooth for marking the shoulder on the face of the spoke. The scribe is placed along the spokes, its foot resting firmly at a point where the spokes enter the nave and the shoulder mark is then scribed on each spoke. In a wheel with a diameter of 4 feet 10 inches, the shoulder will lie 1 foot $7\frac{1}{2}$ inches up the spoke from the hub, i.e. from half the height of the wheel (2 feet 5 inches) deduct half the diameter of the nave (6 inches) and the depth of the felloe ($3\frac{1}{2}$ inches). First of all the front shoulders are cut with a tenon saw and split with chisel and mallet, and then the process is continued for the back shoulder, the saw cuts being continued all the way round the spokes. The tongues are then trimmed to a slightly oval shape; a thin slice of oak or ash with a hole in it is carefully tried over every tongue to make sure that it will fit into the auger hole in the felloes. Through the centre of the tongue a notch is cut, and when the felloes have been fitted on with the end of the spoke appearing through its lower end, a wedge is hammered into the notch to tighten the spoke in the felloe.

## (3)   FELLOES

In Britain early spoked wheels had one-piece ash felloes, steamed to the correct shape and held there by an iron clamp. Although one-piece felloes continued in use alongside the sectional types until the close of the Middle Ages, there is no evidence to suggest that they were used in this country during the last four hundred years. From the sixteenth century onwards the practice has been to allow two spokes for each felloe so that on all wheels one finds an even number of spokes.

More often than not ash is preferred for felloe making due to its great flexibility and strength, but less satisfactory wheels have been made with elm or beech felloes.

The wheelwright buys suitable timber and transports it to his

15   Saw-pit at Pontrilas, Hereford, 1920

yard. In the past a pair of sawyers would visit the yard and cut all the timber required by the wheelwright into required lengths. With the aid of the ring dog, a circular piece of metal to which was attached a strong iron hook, the timber was levered into place over the saw pit. It was held there with a pair of timber dogs, each pair consisting of a staple-like iron bar with a point at each end. One point was tapped into the timber of the saw pit while the other was fixed to the crosswise baulks. Using a pit-saw with a blade of up to ten feet in length, the junior, or bottom, sawyer in the pit and the top sawyer standing on the log above cut the timber into the appropriate lengths needed by the wheelwright. More often than not the felloe blocks, as the untrimmed felloes were called, were the lengths left over after the preparation of longer planks, such as those left over from sawing the side planks or shafts of the wagon.

After sawing, the wheelwright roughly shaped the blocks while still green before stacking them for seasoning. An adze and an axe were generally the only tools required for this. The block was held in a felloe horse, a low bench, and the craftsman, standing astride the horse, would adze out the belly of the block to the correct shape.

Each felloe would be some 30 inches long and some $3\frac{1}{2}$ inches square.

More recently felloes were sawn with a thin bladed frame saw, each felloe conforming to one of the many patterns kept by the wheelwright in his shop. After facing up the felloes with bow and hand saws the pattern is laid on and the felloes marked for any inequality or unevenness. They are planed with the jack plane, the belly trimmed with the adze and the sole with the axe if necessary.

The wheel is laid face downwards on the low wheel stool. With the bevel the right angle is obtained for the ends of the felloes which are cut to the required size. The felloe is then placed on the wheel, resting on two spokes close to the shoulders, and the position of each spoke is marked on the felloe in which holes are bored with a $1\frac{1}{2}$ inch auger. The dowel holes at each end of the felloe are then bored some $2\frac{1}{2}$ inches deep and square to the cut end. As these dowel holes are being bored endways to the grain, it is extremely difficult to keep the point of the auger or twist bit from running with the grain. After all the holes are bored the felloe is trimmed with the drawknife, compass plane and spokeshave. A hard chamfer is taken from the back of the felloe between the spoke holes, for that part of the wheel is regarded as superfluous weight where no great pressure bears. A smaller chamfer, about $\frac{1}{8}$ inch deep, is taken off the front, but great care must be taken not to remove too much off the front of the felloe as it might weaken the spoke shoulders, which bear heavily at that point.

With the felloe complete, the next step is to make the dowels that bind the felloes together and the wedges that close the gap between each felloe. The dowels are $4\frac{1}{2}$ inches long and 1 inch in diameter. They are cleft with an axe out of dry, tough oak, commonly from old spokes and hammered one into the end of each felloe until 2 inches of dowel is left protruding.

The wedges are prepared out of the larger pieces of waste split off the backs of the spokes when making the tongues. They are chopped to the required size of $3\frac{1}{2}$ inches long, $1\frac{1}{2}$ inches wide and $\frac{5}{8}$ inch thick.

With the wheel fully prepared, the next step is to fit the felloes on the spokes. The wheel is placed face downwards on the stool and a felloe tapped in some $\frac{3}{4}$ inch on to one spoke. It will be found that the second spoke is too wide apart to enter the felloe due to the radial character of the wheel which causes the spoke tongues to be wider apart than the shoulders. In order to bring two spokes together so that the felloe can be slipped on easily, a tool called a spoke dog is required. This consists of an ash stick 3 feet or more in length. About 6 inches from the bottom of this handle is a slit which accommodates an iron hook some 20 inches in length. On this iron hook there are eight holes and, by selecting one of these holes and fitting an iron peg through it to form a wrench, the wheelwright is able to adjust the dog to the size of wheel on which he is operating. With the wooden handle against his left shoulder and the other end of the stick behind the already fitted spoke, the iron hook is placed around the spoke that needs drawing. The craftsman presses the handle of the spoke dog forwards and as he does so the two spokes are drawn together so that the felloe can be tapped on. When all the felloes have been fitted in this way wedges are driven into the spaces in between each felloe. All that now remains is to level the projecting wedges with the draw knife and spoke shave. The outer edge of the spoke tongues that appear at the sole of the felloes are gouged out $\frac{1}{8}$ inch below the sole so that they will not press against the tyre when it is fitted. The whole wheel is placed on the wheel stool and finished with jack plane and spokeshave, each joint being tested with a straight edge.

## (4) TYRE

Although hooped tyres were known in Iron Age times, straked wheels were nevertheless very common up to the end of the nineteenth century. It seems surprising that a somewhat primitive method of tyring should have held its own when a more advanced method was already well known. There are a number of reasons for this. In the first place the manufacture of strakes is a far easier process than the manufacture of hoops, and a relatively unskilled man can be entrusted to make and fix strakes. The tyre of a dished wheel is slightly conical in shape to allow for the conical shape of the wheel. It is far easier to curve and cone a strake no more than 2 feet 6 inches long than to curve and cone a hoop with a diameter of 6 feet or more. Secondly, fixing strakes to the wheel is a far easier process and the equipment used is much less elaborate. In clay districts, where broad wheeled vehicles were common, to fit a hooped tyre, 6 inches or broader, would be extremely heavy

(a) Removing the red-hot tyre from the furnace

(b) Lifting the hot tyre on to the wheel

16 Tyring a wheel at Ardington, Berkshire, 1959

work so that it was customary in those districts to shoe wheels with two or even three lines of strakes. In other areas, notably Somerset, it was customary to shoe the wheels with a line of strakes as well as a tyre. On hard hilly roads these vehicles rode on the straked half, which was bigger in circumference than the hooped half due to the bevel of the wheel. On soft soil the vehicles travelled on the

(c) Levering the tyre into position

(d) Pouring on water to shrink the tyre

(e) Plunging the wheel into water

F

whole width of the wheel. It seems that vehicles with straked wheels were preferred in hilly districts since they slipped less when traversing an inclined bank.

Another reason why straked wheels persisted was the fact that the wheelwright conserved fuel in the process of making them. Whereas a large number of strakes could be placed in a furnace to heat, only two hoops at the most could be heated at the same time. An additional advantage of strakes is that in very dry weather, when the whole wheel was liable to shrink, strakes stayed on, whereas a tyre might become loose. Finally, when the strakes were worn out, they could easily be replaced one by one, without removing all of them at once and the farmer himself could do this work. If a tyre became worn, however, it would have to be removed and the whole wheel re-tyred by a skilled craftsman.

The wheel is taken to the blacksmith for tyring. In most country yards both the wheelwright's shop and the blacksmith's forge are located in the same place and many wheelwrights are competent to work both in wood and metal. In the larger yards where two or more craftsmen work, it is customary for each one to specialise in a particular category of work. For example, in a wheelwright's yard at Watlington, Oxfordshire, in 1953, one man was employed as a wood worker, being responsible for making the wooden parts of wheels, undertaking and general carpentry, while another craftsman was responsible for all the metal work, tyring wheels, repairing agricultural machinery and general blacksmithing. For the tyring process at least two men are required.

First of all a bar of metal, some 16 feet long, $2\frac{1}{2}$ inches wide and $\frac{3}{4}$ inch thick from which the tyre will be made is laid flat on the ground. A chalk mark is made on the rim of the measuring wheel or traveller, a disc about 10 inches across with a handle attached to its centre, and another mark is made on the rim of the untyred wheel. With the chalk marks as starting points, the traveller is pushed around the wheel rim and the number of turns noted. It is then run along the bar of iron in order to obtain the correct length of iron required for the tyre. Allowance has to be made, however, for 'shutting' or welding the loose ends of the tyre, and for the expansion of the tyre when heated. Generally for a wheel 5 feet in diameter, the diameter of the tyre should be approximately $1\frac{5}{8}$ inches less than the wheel. As soon as the correct length is ascertained, the bar is cut off at the chalk mark. One man holds a cold chisel over the iron as it rests on the anvil, while another smites it with a sledge hammer.

The next operation is known as 'scarfing down', which consists of flattening each end of the bar with a sledge hammer Each is first of all heated, hammered, and a hole punched through each scarf. The iron is then passed through the rollers of the tyre bender, to the required shape. In the older country workshops the trye was bent by hand on a post bender—a difficult and intricate process.

The next stage is the welding of the two loose ends of the tyre to form a complete hoop. By straining the two ends together, a nail is thrust through the holes that have been punched in the scarfs and the joint welded. After welding, two or three nail holes are punched in the tyre and all is now ready for the process of tyring the wheel.

An open fire of wood shavings, straw or peat is made in the corner of the yard and the tyre placed on them. In some wheelwright's yards a special type of upright brick oven, the furnace, is used for heating the tyre, and within the oven one or two tyres can be heated at the same time. Meanwhile the untyred wheel is screwed face down on a heavy iron tyring platform—a permanent feature in a wheelwright's yard—by means of a threaded rod passing through a ring below the central hole of the platform, which passes through the hub of the wheel. The wheel is fixed rigidly with the nave resting in the central hole and the spokes and felloes supported along their whole length by the face of the platform. All is now ready for the tyring process. Two or three men with long handled tyring tongs grasp the red hot tyre from the fire and throw it down on the ground, so that all pieces of fuel and rubbish adhering to it are knocked off. It is grasped again, dropped into position on the wheel and levered with tyre dogs and beaten with sledge hammers until it is in place. Water is then poured on the rim, and as the tyre shrinks the wheel is tightened under the enormous pressure of contraction.

The amount of dish on a wheel can be controlled by adjusting the central screw of the tyring platform. Since the wheel is placed on the platform face downwards, that is with the convex side upwards, the effect of the tyre is to pronounce the dish. By loosening the screw of the central rod, the hub set free from pressure rises up an inch or more, and the contracting tyre tightens still further to increase the dish. If little dish is required beyond what

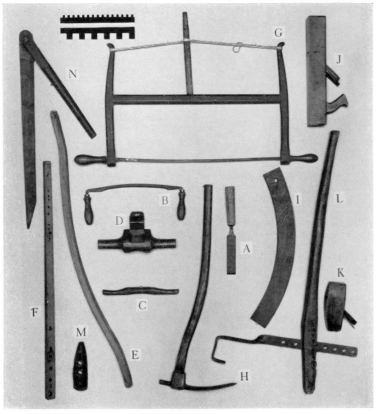

| | | | |
|---|---|---|---|
| A | Turning chisel | H | Adze |
| B | Drawknife | I | Felloe pattern |
| C | Spokeshave | J | Smoothing plane |
| D | Jarvis for spoke making | K | Jack plane |
| E | Crooked stick, or bridle, for fitting felloes | L | Spoke dog |
| F | Spoke set gauge | M | Spoke tongue gauge |
| G | Bow saw for cutting felloes | N | Bevel |

17 (a) The wheelwright's tools

has already been made on the wheel, then the wheel is kept tightly in place on the platform until the whole tyre cools and the contraction ceases.

The completed wheel is then taken from the tyring platform and trundled to the shoeing hole. This narrow pit, always kept full of water, is some 5 feet long and has an iron bar at its back.

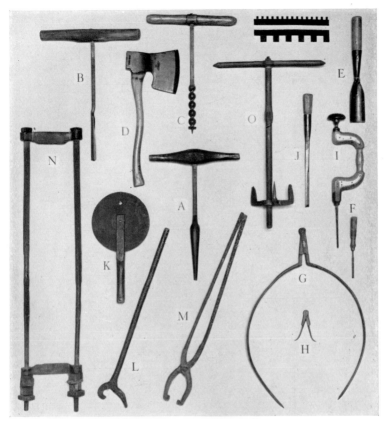

| | | | | |
|---|---|---|---|---|
| A | Spoon auger | I | Brace and bit |
| B | Shell auger | J | Morticing chisel, or bruzz |
| C | Spiral auger | K | Measuring wheel, or traveller |
| D | Axe | L | Tyre dog |
| E | Gouge | M | Tyre tongs |
| F | Heading chisel | N | Samson, used in straking |
| G | Inside callipers | O | Boxing engine |
| H | Outside callipers | | |

17  (b) The wheelwright's tools

The wheel is placed on this bar and swung around until the tyre is perfectly cool. In the meantime it is checked with gauges and the felloe joints are hammered into the correct position if necessary, before the tyre cools. The whole wheel is then finished off on the wheel stool, or is hung on a wheel horse and cleaned with spokeshave, pieces of broken glass and sand paper. The great force of the contracting tyre may have frayed some of the wood, especially along the felloe joints, and the frayed ends have to be cleaned off. Scorch marks from the red hot tyre are scraped away with pieces of glass, and a smooth surface for the paint made with sandpaper. Finally, three or more nails are knocked into the tyre,

17   (c) Wheelwright's lathe

and although this is not common practice throughout the country, the nails ensure that the tyres keep in position if they become loose in dry weather.

When the wider type of wheel requiring a tyre 3 inches or more in width is shod, the process is slightly different. The wheel is placed face upwards on the tyring platform, and as the dished spokes and the felloes of the wheel lay some inches from the surface of the platform, they have to be supported with blocks of wood. The heated tyre is picked out of the fire with long hooks, it is turned with its larger diameter downwards, dropped over the wheel and hammered into position.

The process of straking a wheel is completely different from that of tyring it with a hoop. A strake may be defined as an iron shoe fitted over the junction of two felloes, stretching from the centre of one felloe to the centre of the next. Strakes are cut off from a cold iron bar with sledge hammer and hard chisels. They are then heated, hammered to a convex shape over the anvil, and four or more square nail holes are punched near the end of each strake.

Nails for fitting the strakes to the wheel are generally large, blunt and tapered, with large square heads. These are made by hand, a process that usually requires the services of two men. A rod of iron is heated and hammered to the required shape on the anvil. In the anvil orifice a cutting tool called a hardy is inserted, and the rod is cut to the required size on this tool. The hot nail is then inserted in the nail heading tool and the head made. The heading tool consists of a flat metal bar finished at one or both ends with a deep perforated knob; the perforation complying with the size and shape of the nail shank. The perforation is also countersunk to correspond with the nail head. After inserting the red hot piece of metal rod in the nail hole, the tool is placed over the anvil orifice and the head hammered into shape.

Before fitting the strakes each felloe has to be bored since the nails are too large and blunt to enter the hardwood of the felloes without preparatory boring. Great care must be taken therefore to see that the nail holes in the strake correspond to those in the felloe.

The strakes are placed in the fire of the strake chimney, a closed-in brick fireplace a little distance from the shoeing hole. The wheel itself is hung through the hub to the low post at the back of the hold, its bottom resting in water. Nearby, in a bucketful of water, are the long square nails, while close at hand is the heavy cramping gear known as a samson, which will be used to bring the ends of the adjacent spokes as close together as possible, so that the strakes can be nailed on. A nail is tapped into the side of the rim to make the position of the first strake on the wheel. Seizing the red hot strake with a pair of long-handled tongs, the wheelwright rushes over to the wheel and places it in the correct position. Nails are knocked in at each end and, before the wheel catches fire, it is turned around so that the newly fitted strake lies in the water at the bottom of the hole. The cooling causes contraction and the whole wheel is tightened. The second strake to be fitted on the felloes is fixed directly opposite to the first one. This is done in order to distribute the application of the binding strain as equally as possible around the wheel. The whole process is continued until the entire wheel is shod. One nail in each strake fitted is driven in firmly but not driven home. The reason for this is that when the whole wheel is shod the felloe joints have to be brought closer together and this is done with the samson. One end of the samson is hooked around a spoke, while the other end is hooked over the projecting nail. The nuts on the samson are then gradually tightened until the intervening felloe joints are close together. The process is repeated for every joint around the wheel. All the nails are then driven home and the wheel is complete.

## (5)  BOX

The final process in the manufacture of a wheel consists of fixing a cast-iron bearing or box in the centre of the hub to take the wear of the revolving wheel. Although in some parts of the country up to fairly recent times vehicles were equipped with disc wheels that turned in one piece with the axles, the use of some kind of bearing has been known for many centuries. The usual method of boxing in the nineteenth century was to fix a pair of cast-iron bushes inside the nave and this method remained in vogue for as long as wooden axle arms were made. Since wheels of post-sixteenth-century date were greatly dished and accommodated a tapered axle arm, a pair of semi-cone-shaped bushes had to be fitted. These were tapered according to the the taper of the axle arm. When iron axle arms became common, a rather different type of cast-iron box was required. This merely consisted of a heavy cone-shaped iron pipe, running the whole length of the nave, and fitted tightly into the centre of the nave.

A large tapering hole is cut through the centre of the nave with heavy boxing chisels and gouges. In the past the central auger hole was first of all enlarged with the V-shaped morticing chisel and shaped with a wide gouge. More recently a tool known as a boxing engine was used. This engine consists of a round bar of iron 2 feet or 3 feet long, screwed down half its length with three pronged grips to hold the nave. Half way up the stem there is a small cutter which can be adjusted according to the width of boring required. The boring bar is passed through the centre of the hub and is turned by the cross handle at the top, and while the three prongs hold the nave steady the cutter bores out the hole to the required size. The hole itself is made a little too large for the box, for it is extremely important to have the pivoting true and this can only be done by trial and error.

The axle arm is temporarily fitted to the bench and the wheel, with the box in position, is hung on it so that it swings just clear of the ground. A small block of wood is then placed on the floor underneath the wheel just touching the edge of the tyre.

Slowly the wheel is turned around and, as it swings clear of the block, small wedges of heart of oak are hammered into the end grain of the hub. The process is continued until the box is firmly wedged and centred in the hub. The iron hub bonds prevent the elm of the hub from spreading outwards, and only by spreading inwards can it make room for the wedges. It is important that the wedges themselves do not touch the brittle cast-iron box, and for this reason they are driven into the elm itself, a place for them being started with a broad iron wedge. As soon as the box is made immovable, all that remains to be done is to chisel off the ends of the wedges and smooth them with a spoke shave. The wedges are so tightly fitted that only the difference between the grain of elm and oak shows their position in the finished wheel.

Last of all, a block of wood at the outer end of the hub is cut out in order that the linch pin holding the wheel to the axle arm can be removed when required. In the past this chock of wood was lined with a little straw or sacking so that it kept is position on the moving vehicle. A clasp and staple are fitted to the hub in order to prevent the chock of wood from falling outwards.

## 2 UNDERCARRIAGE

### (1) AXLES

The old type of axle consisted of a length of well-seasoned beech, but in areas where beech was scarce, as in parts of East Anglia, ash was used as an alternative. The average length of the axle was 6 feet 8 inches. The ends were tapered and rounded to form the axle arms. Each arm equalled the length of the wheel nave and varied in diameter from $5\frac{1}{2}$ inches at the bed to $3\frac{1}{2}$ inches at the extremity.

The axle arm was given a distinct downward pitch so that the lower, loaded spoke of the dished wheel, which momentarily took the weight of the loaded vehicle, would be in a vertical position.

In addition to the downward pitch each arm also had a distinct forward cant, so that the wheels tended to run towards the central line of the wagon. While the downward pitch tended to draw the wheels outwards away from the axle tree and against the linch-pin, the forward cant had the directly opposite effect. It was the wheelwright's aim to balance these two opposing forces.[4] In some parts of the country a well balanced wheel of this type was known

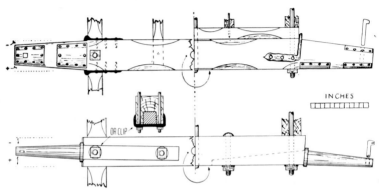

18   Wooden axle with wooden arms (above) and wooden axle with iron arms (below)

as 'a pitch and gather wheel', while in others it was described as 'dished and forehung'.

The underside of the axle arm, made horizontal and in the same plane as the underside of the axle bed, was shod with a very smooth plate of iron or steel. This clout plate lay flush with the wood and was fastened to the arm by six or more flat-headed, countersunk nails. Another clout plate 4 inches long and 3 inches wide was attached to the top slot of the arm encircling the hole where the linch-pin was fitted. Finally a circular or semi-circular iron collar was fitted around the inner edge of each arm, in order to prevent the wheel from wearing out the shoulder of the axle tree as it knocked against it with the lateral movement of the vehicle.

The process of making a wooden axle is extremely complicated. Not only must the square axle bed be sawn in such a way as to produce the maximum strength but, in addition, great care must be taken in shaping the arms so that the correct balance between dish and foreway is struck.

Many wheelwrights, such at Sturt,[5] set out axles with little preparatory measurements, depending entirely on craft instinct and eyesight for the correct alignment of axle arms. Others, such as the late George Weller, who practised his craft in the Sussex village of Sompting until his death in 1953, used very careful measurements and an intricate full-scale pattern.[6]

As an alternative to the all-wooden axle, many wagons were later fitted with iron axle arms on a wooden axle bed; a method of construction widely practised even in the early years of the last century. The axle bed of well seasoned beech was cut to the approximate width of the wagon frame and two deep grooves were cut on its under side to take the square sections of the axle arms. Each groove was cut at a slight angle, so that when the iron arm was fitted on, the square portion being bedded in the axle tree, the round portion which carried the wheel pointed downwards and also slightly forwards, that is, it was dished and forehung. Finally, with draw knife and spoke shave the axle bed was lightly shaved wherever possible, but great care had to be taken not to weaken it.

Although all-wooden axles and axles of wood with iron arms persisted as long as the distinctly regional types of farm wagon persisted, all-iron axles were nevertheless fitted to wagons from as early as 1840. The boat wagons of the late nineteenth or early twentieth centuries were fitted with iron axles, and it was customary to fit these in a groove cut on the underside of a wooden beam.

In order to fit the wheel to the axle arm, an iron linch-pin is passed through the end of the hub, penetrating the hole bored in the end of the axle arm. For a wooden axle arm, the linch-pin is square in cross-section, and is made from a $\frac{3}{4}$ inch bar. An iron axle arm on the other hand has an oblong linch-pin, measuring 1 inch by $\frac{5}{8}$ inch.

In some cases the top of the linch-pin fitted to the rear wheel of a wagon is hammered out and pierced through the centre to produce a ring. When the vehicle is taken up-hill the hooked end of the roller scotch chain is passed through this ring.

## (2) FORE AND REAR CARRIAGES

The front axle of a wagon is capped by a heavy block of timber 4 or 5 inches square which is known as a bolster (Fig. 19). This may be of beech, ash or oak and is the same width as the axle bed itself. Sometimes the ends of the bolster are tapered and chamfered so that they overlap the inner side of the nave.

Between the axle and the bolster a pair of outside hounds is found. Each hound, made of oak or ash, is $3\frac{1}{2}$ inches or 4 inches square and projects some 30 inches beyond the bolster. Behind the bolster the hounds taper and are bolted to the horizontal, slightly curved slider bar. The slider, which may be of beech, ash or oak,

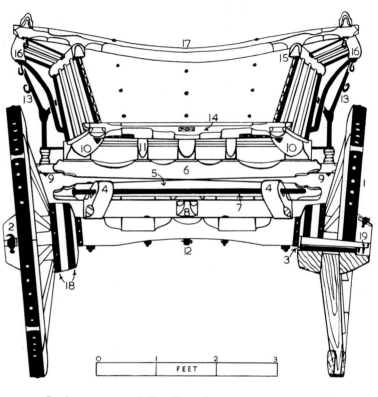

| | | |
|---|---|---|
| 1 Strake | 8 Coupling pole | 15 Inner top rail |
| 2 Nave | 9 Main cross-bar | 16 Outer top rail |
| 3 Axle | 10 Side frame | 17 Front top rail |
| 4 Hounds | 11 Summer | 18 Nave bonds |
| 5 Bolster | 12 King pin | 19 Linch pin |
| 6 Pillow | 13 Side support | |
| 7 Shaft bar | 14 Forebridge | |

19   Front elevation of a Bow Wagon

carries a wearing plate of iron on its upper side. This, together with a plate on the underside of the pole that couples the fore- and rear carriages of the vehicle takes the wear as the wagon is pulled along and turned.

In addition to a pair of outside hounds, many forecarriages are equipped with two inside hounds which end just behind the bolster. On some wagons, however, especially on those used in hilly districts, the inside hounds continue parallel to the outside hounds and are bolted to the slider bar.

In front of the bolster the hounds may be joined together by one or two oak slats known as hound shutters. The shafts which are fitted to the front of the hounds may be attached to the forecarriage in one of two ways. Either a single pair is attached by a rod pin passing through the ends of the shafts and hounds, or the hounds are morticed to a square length of wood known as a splinter bar, especially when the wagon is designed to be pulled by two horses abreast. Two sets of barrel eyes are fitted to this bar, and the shafts are attached to it by passing two iron pins through each set of barrel eyes, one set on the shafts, the other on the splinter bar. Generally this arrangement can be adapted for one or two pairs of shafts.

The axle, bolster, hounds and slider constitute the forecarriage of a wagon. Above the bolster, and exactly the same length, is another member bolted to the bottom of the wagon known as a pillow. This is made of beech, ash or oak. To hold the body of the wagon to the undercarriage, a $1\frac{1}{2}$ inch hole is bored through the floor of the vehicle, the pillow, bolster, front end of the coupling pole and the axle. Great care must be taken to ensure that this hole coincides vertically. Through this hole passes the important king pin which links not only the body of the vehicle and its undercarriage, but also the fore carriage and the rear carriage.

The rear carriage is composed of an axle tree and a bolster of equal size. In general there is no pillow at the back and the body rests in grooves cut in the bolster, held in place by two vertical clips fixed to its ends. The body of the wagon may therefore be regarded merely as resting on the undercarriage, since its only rigid connection is the king pin which runs through the floor and forecarriage. Indeed, in some parts of the country, notably West Sussex, the whole body can be removed and the undercarriage used as a timber carriage. As far as we know, Constable's famous

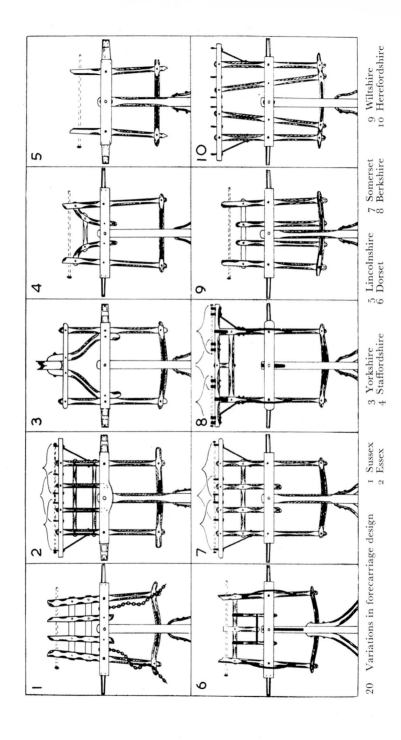

20 Variations in forecarriage design

1 Sussex
2 Essex
3 Yorkshire
4 Staffordshire
5 Lincolnshire
6 Dorset
7 Somerset
8 Berkshire
9 Wiltshire
10 Herefordshire

painting *The Hay Wain* shows this type, for the body seems merely to rest on an undercarriage which, like the ordinary timber carriage, has a long pole projecting at the back. The length of vehicle could be varied according to the length of timber carried.

The tension set up by the pull of the horses is transmitted from the forecarriage to the rear carriage not along the frame of the wagon, but along the coupling pole. This ash pole extends from the king pin, between the bolster and front axle, to the rear carriage, where it passes again between the bolster and rear axle. In some cases this pole is straight, in others it curves over the slider of the forecarriage. Sometimes, too, it ends in a chamfer just behind the rear axle. Sometimes it continues upwards and is fixed to the rear cross-bar of the body. This latter arrangement helps to support the rear end of the wagon and is the only attachment between body and carriage in this part of the vehicle.

In the rear carriage, in place of the hounds of the forecarriage, the pole braces pass through the space between bolster and axle tree. These two braces, projecting 4 or 5 inches beyond the bolster, run forward at an angle to meet the coupling pole about half way along its length. The main purpose of this attachment is to keep the pole square with the main part of the rear carriage.

In the fore- and rear carriage the whole structure is made fast by two long bolts passing through the bolster and axle tree of each carriage. The heads of the bolts are embedded in the bolster. In the forecarriage they pass through the bolster, the outside hounds and axle tree, and are fastened at the bottom by square heavy nuts. In the rear carriage the bolts pass through the bolster, the forked ends of the coupling pole attachment and the axle tree, and are again secured with a pair of heavy nuts.

## 3 SHAFTS

Wagon shafts are generally made of ash, though in the past elm and oak were occasionally used. When a load of timber had been delivered to the wheelwright's yard, it was customary for the craftsman to assign the various pieces to particular parts of the wagon. Thus, gently curving beech was immediately reserved for felloe making, while straight oak butts were reserved for spoke making. In the same way naturally curved ash butts were immediately reserved for making shafts. Sturt noted that if 'a butt of

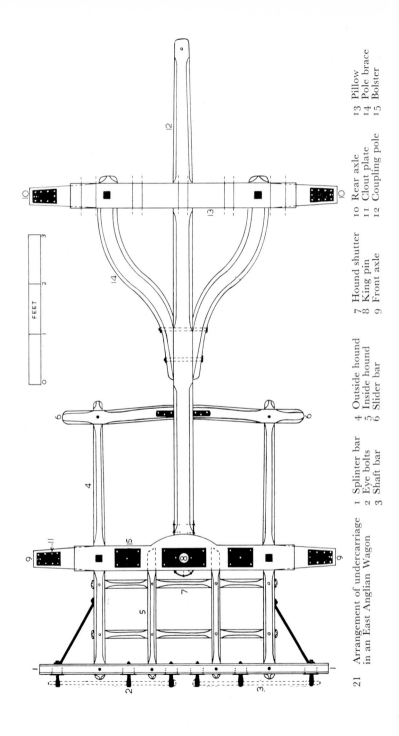

21 Arrangement of undercarriage in an East Anglian Wagon

| 1 Splinter bar | 4 Outside hound | 7 Hound shutter | 10 Rear axle | 13 Pillow |
| 2 Eye bolts | 5 Inside hound | 8 King pin | 11 Clout plate | 14 Pole brace |
| 3 Shaft bar | 6 Slider bar | 9 Front axle | 12 Coupling pole | 15 Bolster |

FEET

ash had the shape of a waggon shaft, it was marked off the right length for that.'[7] The butt was then sawn from end to end so that the two curved pieces provided wood for a pair of shafts. Accurately chalking-in the shape of the shafts, the craftsman using hand saw, drawknife and spokeshave, carefully cuts the shafts to the correct shape and size. Shafts must curve in at the front until they almost touch the horse's collar. At the back more width is required to allow for the curvature of the horse's body and for the swaying action as it moves along.

On some wagons the shafts are attached directly to the hounds of the vehicle's forecarriage by means of a round iron pin $1\frac{1}{4}$ inches thick. In this case the shafts are made some 15 inches longer than usual and are very strongly joined together. The join is made with two curving ash cross-bars set back to back, and firmly morticed into each shaft. In many cases these bars are overlaid with a $\frac{3}{8}$ inch curved iron plating to give added strength. This type of single shaft attachment is generally found on the smaller wagons designed to be drawn by one horse.

On other wagons, especially the larger vehicles where two horses abreast are required, two pairs of shafts are fitted to the splinter bar of the forecarriage. In this case the shafts are rather shorter than those fitted to single shafted wagons and each pair is equipped with a number of eye-bolts. These eye-bolts alternate with other bolts on the splinter bar. The shafts are fastened by means of a $\frac{3}{4}$ inch iron bar that passes through each set of eyes. Wagons which possess splinter bars may be adapted for drawing by one horse or two horses in line abreast. When using the central eye-bolts only a single pair of shafts may be attached to the vehicle.

# 4  BODY

## (1)  FRAMEWORK

From the constructional point of view, the simplest and strongest frame consists of two parallel baulks of oak joined together by a number of cross-pieces. This method of construction, found on all wagons on the European continent, is found too on a number of English farm wagons such as those of Surrey and Wiltshire. The frame consists of two straight side-frames of oak joined together at the front by a straight or slightly curving piece of timber known

G

as a forebridge or front cross-bar. Near the centre of the side frame a heavy cross-bar of well seasoned oak extends at right angles to the sides. At the back, the rear cross-bar, slightly lighter than the main cross-bar, joins the side frames.

Running parallel to the heavy side frames, two lighter members, known as summers, join the forebridge to the rear cross-bar. These summers are either morticed into the three cross-bars or fitted into grooves above or below them.

This rectangular framework, although strong and simply constructed, has one serious drawback in that the fore-wheels soon rub against the straight frame of the wagon on turning. To overcome this disadvantage a number of expedients are adopted. The simplest means is to make the side frames curve upwards at the front, as on the wagons of Shropshire. Since the wheels are dished, on locking they rub against the frame high up the spokes rather than near the nave, as would be the case if the frames were made perfectly horizontal. Where the wheels are intended to rub against the frames, it is customary to fit an iron wearing plate on the frame to prevent rapid wear and the weakening of the timber.

Another method of increasing lock is the groove or notch cut in the side timbers to allow the turning wheels to lock a little beneath the body of the vehicle. This device, which is common in Leicestershire, Rutland and Northamptonshire, would seriously weaken the side frames but for the very thick heavy timber used on the frame.

An expensive method of increasing lock, which was limited to the later carriers' wagons of the eighteenth century, was to carve the side frames into an arch at a point where the fore-wheels would rub. The turning wheels would enter this arch and swing underneath the body until they touched the coupling pole. This notched frame construction is clearly seen on a road wagon at the Museum of English Rural Life. Each side frame consists of a heavy block of timber 6 inches square, sawn into an arch where the fore-wheels go beneath the body. To cut down the weight of such heavy timbers, the frame is greatly chamfered along its outer and lower edges.

By far the most common method of increasing the lock of a vehicle is to make the side frames in two sections so that a waist is provided for the fore-wheels to turn in (Fig. 22). By making the forward side frames with curved sides, the builder sacrifices the ideal form of framing, the simple rectangle, in order to achieve greater lock. In Sussex, for example, the wagons are constructed with a centre cross-bar of oak, measuring 5 feet 6 inches long, 4 inches wide and 6½ inches deep. This is morticed to take both the forward and rear side frames, as well as the pair of summers that run directly from the forebridge to the rear cross-bar. The forward side frame consists of naturally curving oak some 4 inches thick, which is morticed to the cross-bar, and tapered to protrude about 4 inches beyond the cross-bar. The waist made by the forward side frames is about 9 inches deep on either side.

For the floor of the wagon two methods of laying the boards must be noted. Firstly, the vehicle may be cross-boarded; that is, it may have floor boards at right angles to the side frames, running across the wagon. Secondly, a wagon may be long-boarded, with slats running parallel to the side frames. Elm is generally preferred for floorboards since it does not split easily. In fact, it is sometimes said that the knottier the elm, the better it is, since the knots decrease its tendency to split.

For cross-boarding, the elm slats, each one no more than 2 inches thick, are nailed directly across the top of the side frames and summers. For long-boarding a series of keys thinner and lighter than the summers are inserted at right angles to them.

Whereas cross-boarding is perfectly efficient in wagons designed for harvest only as, for instance, the wagons of East Anglia, it is less satisfactory in wagons designed to carry other materials such as gravel, sand or manure. As these must be shovelled out of the wagon, the carter needs a smooth surface for his shovel to slide along. 'One board a shade thicker than the one next to it,' says Sturt,[8] 'will lift a low but exasperating ridge all across the cart, or stones may get wedged in between two cross-boards; or a small stone may force one of the boards up a little; or the edge of an elm board may curl up right in the way of the shovel . . . the shovel should slither along the cart or wagon floor (with long boarding) from front to back as if the floor were greased.'

## (2) SIDES

For the side planks of a wagon a variety of timber can be used, but by far the most popular is ash. In some areas, especially where the wagons are waisted, it is customary to use two different types of timber in the sides. In Sussex, for example, the sharply curving

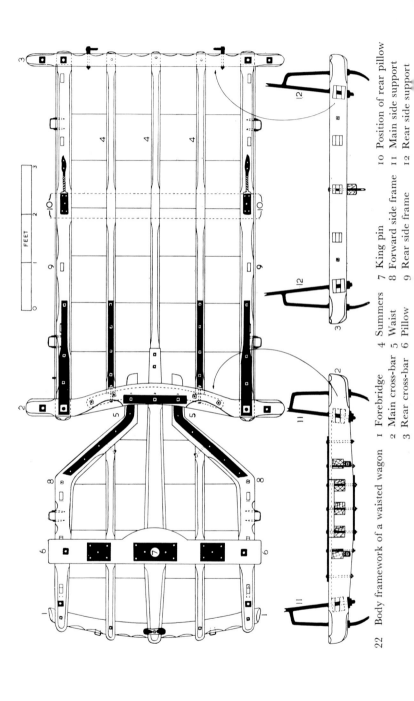

22 Body framework of a waisted wagon

| | | |
|---|---|---|
| 1 Forebridge | 4 Summers | 7 King pin |
| 2 Main cross-bar | 5 Waist | 8 Forward side frame |
| 3 Rear cross-bar | 6 Pillow | 9 Rear side frame |

10 Position of rear pillow
11 Main side support
12 Rear side support

sides in front of the main cross-bar are made of non-splitting poplar, while to the back of the centre bar the straight sides are made of ash. While a variety of timber can be used for the side planks, the spindles or standards supporting the sides are invariably ash. The top rail of the sides, corresponding to the main frame of the vehicle, is made of oak when the top is perfectly flat from end to end, as in Hertfordshire wagons. Where the top rail is curved, as in the East Anglian wagons, or bow shaped, as in West Country wagons, ash is generally preferred because of its pliable character.

As has been said previously, British wagons are characterised by three methods of side construction. They may be spindle-sided, panel-sided or plank-sided. Although the general method of construction is similar in all three types, the method of assembly does vary slightly.

Where a spindle-sided wagon is shallow-bodied, for example the Cotswold wagon (Fig. 23a), then the spindles run directly from the outer end of the side frame to the inner top-rail with no intervening mid-rail. The lower ends of the spindles are first of all morticed into the side frames; the single ash plank is then put in place; finally the curving inner top-rail is morticed on the other end of the spindles. The spindles are nailed or pegged to the side planks.

While one width of plank is generally large enough to board each side of a shallow-bodied wagon, two or more planks are required for the deeper-bodied vehicles, such as those of Lincolnshire and the East Midlands (Fig. 23b). The stages in the building up of a deep-bodied, spindle-sided vehicle are as follows: a number of ash spindles, each approximately 2 feet long, are morticed into the side frame of the wagon, and their top ends cut according to the curve of the top-rail; an ash plank is then clamped on to the frame and fitted tightly against the spindles; an ash or oak mid-rail measuring some 3 inches square, is now clamped down on the lower plank, its curve following that of the top of the plank already fitted. The spindles pass through holes already cut in this mid-rail. After this has been secured, the top plank is attached, followed by the inner top-rail. When the spindles have been morticed into this top-rail, the sides of the wagon are complete. In some cases, where the body is very deep, such as in Leicestershire wagons, two mid-rails are required. A typical Leicestershire wagon would have the following parts to its sides, the side frame, into which are morticed 30 or more spindles; above this the lower side plank, followed by the lower mid-rail; next is the middle side plank, the upper mid-rail, the top side plank and finally the inner top-rail, into which the ends of the spindles are morticed.

In panel-sided vehicles (Figs. 23c, 23d) the method of making and assembling the sides is the same as for spindle-sided vehicles. Because the flat ash standards are less easily bent than the rounded spindles however, the technique for deeper bodied vehicles is slightly different. For example, in Essex the procedure is as follows: the upright ash standards are first morticed into the side frame; then the lower side plank fitted. The slats to the rear of the waist are almost vertical, and can be pegged in at an early stage of construction. But since the wagon has a high frontboard, the standards in front of the waist leave the frame at different angles. Hence they cannot be finally pegged into the frame until the mid-rail is in place. The three fore-standards on the Essex wagon are fitted to the mid-rail while it is on the ground, the craftsman finding the correct alignment by trial and error. Once the mortices for the six vertical rear standards have been bored in the mid-rail, the mid-rail and fore-standards are clamped into place on the lower plank. This is followed by the upper side plank and the inner top-rail of sharply curving ash.

In plank-sided wagons (Fig. 23e) the various planks are morticed into one another and two or three iron bars screwed on to keep them in place. In general construction plank-sided wagons are extremely simple and can be built very quickly.

## (3) SIDE SUPPORTS

A wagon is equipped with a number of wooden or iron side supports or strouters. The function of these supports is twofold. Firstly, they act as a buttress for the side planks of a vehicle. Secondly, in many cases, they provide a support for the overhanging sideboards.

The fore-part of the wagon sides is generally nailed or morticed to the frontboard, but at the back there is no such support. Since the tailboard is removable and does not provide any support for the body when open, some kind of bracing is required. In the centre, too, some form of buttress is needed to prevent the body spreading outwards. With heavy loads of hay or other material, the strain on the sides of the vehicle is great and buttresses are

(a) Spindle-sided

(b) Spindle-sided with midrail

(c) Panel-sided

(d) Panel-sided with midrail

(e) Plank-sided

23   Methods of side construction

required to prevent the side planks from spreading. The two strongest supports are therefore at the back of the vehicle, to prevent lateral spread at the tail, while an equally stout support is found at the centre cross-bar. In addition, there are a number of intermediate supports along the length of the body. These are lighter and of more simple design than the tail and centre supports.

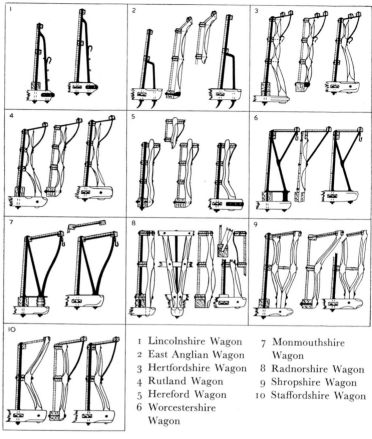

1 Lincolnshire Wagon
2 East Anglian Wagon
3 Hertfordshire Wagon
4 Rutland Wagon
5 Hereford Wagon
6 Worcestershire Wagon
7 Monmouthshire Wagon
8 Radnorshire Wagon
9 Shropshire Wagon
10 Staffordshire Wagon

24   Designs of side supports

Side supports vary from one region to another and may be either of wood or iron. Ash is generally preferred and often the supports are cut from an old pair of shafts. At the bottom they are either morticed to the cross-bars or bolted to the frame of the wagon. In some cases they may rest in an iron socket nailed to the frame. Iron supports on the other hand are morticed to the centre and rear cross-bar, and the sides of the wagon built up from the front with the supports already in place.

### (4)  SIDEBOARDS

In order to increase the loading capacity of a vehicle, it is

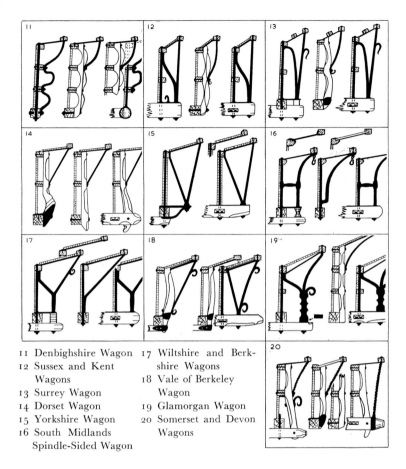

11 Denbighshire Wagon
12 Sussex and Kent Wagons
13 Surrey Wagon
14 Dorset Wagon
15 Yorkshire Wagon
16 South Midlands Spindle-Sided Wagon
17 Wiltshire and Berkshire Wagons
18 Vale of Berkeley Wagon
19 Glamorgan Wagon
20 Somerset and Devon Wagons

customary not only to fit fore- and tail ladders, but also overhanging sideboards permanently attached to the top of the wagon. These boards vary in width and construction from one part of the country to the other, ranging from the very wide, almost horizontal sideboards of Wiltshire wagons to the narrow, almost vertical boards of East Anglian wagons.

There is considerable confusion in the terminology of these overhanging boards. They are known in various parts of the country as raves, raths, rails, lades, lade boards, everings, shelvings and surboards. For this reason the purely descriptive word 'sideboards' has been adopted.

It is interesting to note that sideboards do not occur in any continental wagons, and the feature is peculiar to English farm vehicles. Prior to the mid-eighteenth century they were not found on English wagons, although they occur sometimes on the traditional two-wheeled tumbrils. The old carriers' wagons designed to carry goods in sacks, bales or boxes, did not require sideboards. But in the eighteenth century when country craftsmen adopted the design of the old stage wagons for the new but lighter farm wagons they tried to increase the loading space available. The introduction of sideboards at this period raised the efficiency of the farm wagon—a vehicle designed primarily for harvest work and carrying light but bulky materials such as corn, hay or straw. Running parallel with the inner top-rail of the wagon body is another rail, known as the outer top-rail. The distance between these top-rails varies according to the width of sideboard required. The outer top-rail is supported by a series of sideboard supports, either of wood or iron. These run from the rail to the body of the wagon, or are connected to the main or intermediate side supports of the vehicle. In many cases the sideboard supports form part of the actual side supports of the vehicle. The outer top rail is also connected to the inner rail by a series of iron brackets or by wooden standards morticed into each member.

The method of filling in the space between these top-rails varies from one part of the country to the other. Firstly the space may be left open, as in Bedfordshire wagons, where the two rails are fairly close together. Secondly, an intermediate rail of ash or oak may run in between the top rails and parallel to them as in Surrey wagons. Thirdly, the space may be filled in with one or more elm slats, so as to make the sideboards solid. Many wagons, including the Wiltshire wagon, have this arrangement. Lastly, if the weight of the vehicle must be cut down, to retain wide sideboards, a series of spindles joins the two rails together. The Cotswold wagon has this arrangement.

In some cases, notably in Lincolnshire wagons, the sideboards are removable. These, which are a combination of fore- and tail ladders and sideboards, will be discussed later.

## (5)   END BOARDS

In most cases the frontboards of wagons are permanently fastened to the framework and are morticed into the sides of the wagon.

In East Anglia, however, frontboards are removable and are only held to the sides by a pair of pins. In that region the frontboard is made in two sections, one over the other, and one or both can be removed and replaced by a foreladder projecting over the horse's back.

The most common type of frontboard consists of the following sections: a front top-rail morticed to the front ends of the outer top-rail; an inner top-rail fitting into a groove cut in the front top-rail; at the bottom the forebridge, attached to the main side timbers and summers. A solid plank is generally used to fill the rectangular space formed by the inner top-rails, the forebridge and the sides of the wagon, but in the case of a deep bodied vehicle two flat planks are required and perhaps a front mid-rail corresponding to the side mid-rails.

The tailboard, on the other hand, is usually removable. An adjusting chain attached to the inner side of the wagon and to the board makes it possible to vary the distance that it can be opened. The tailboard is held to the rear cross-bar, either by a hinge or by an iron standard that fits into a pair of staples on the cross-bar. The iron standards are screwed to the tailboard itself and provide an extra support for the slats of the board.

## 5   METHODS OF BRAKING

### (1)   ROLLER SCOTCHES

The roller scotch is a small cylinder hung on chains behind a wagon wheel when moving up hill, to be in place for scotching a wheel immediately if the wagon stops. One of the chains is fixed to the rear axle bed, or to the back of the wagon frame, the other is hooked through the eye in the linch pin of the rear wheel, otherwise it embraces the whole wheel, or is hooked to an eye in the side frame, just in front of the rear wheel. Rollers are made of tough resistant elm bound with two or more iron bonds and measure some 9 inches in length and 5 inches in diameter. When not in use, the roller is hung on the back axle of the wagon.

### (2)   DOG STICKS

As roller scotches are never fitted to two-wheeled carts, a dog stick is fitted to the rear of the axle bed. This method of braking is also found on many four-wheeled wagons instead of the roller scotch. The dog stick consists of a wooden stick, tipped with an iron point. When the wagon goes uphill the pointed end of the stick drags behind. If the wagon stops the metal point digs into the ground and prevents the vehicle from slipping backwards. The stick, of ash some 48 inches long, is fitted to the rear axle. When not in use the tipped end is pushed through an iron ring or socket on the rear cross-bar.

## (3)  DRAG SHOES

The drag shoe, known also in various parts of the country as a drug shoe, skid pin, drug bat or drag bat, is a thick cast iron shoe hanging from the wagon by a stout chain just in front of one of the rear wheels. For going downhill the shoe is slipped under one of the hind wheels so that it skids and will not rotate. This restricts the pace of the vehicle. The wear and tear of dragging on the road is taken by the convex sole of the shoe and not the tyre that lies on it. Drag shoes are about 12 inches long but their width varies according to the width of the tyre of the wagon.

In some places, notably Dorset and Devon, where the wagons are small and not equipped with drag shoes, it is customary to chain one of the rear wheels to the body frame, thus locking it. 'Instead of a cast iron slide for the wheel to rest upon, the tyre is made $1\frac{1}{2}$ inches thicker in one place, which place being made to slide along the ground is soon reduced to the same thickness as the rest.'[9]

In West Wales another device for braking is used. This is a convex iron shoe, 2 feet long, which is in effect a section of a tyre. It has two pairs of flanges, into which the wheel fits and is equipped with a pair of small iron wheels, no more than 9 inches in diameter. For going downhill the shoe is placed under one of the wheels of the vehicle, the small cast iron wheels to the front. It is customary to leave four or five sets of these shoes at the top of a hill for, unlike drag shoes, wheeled shoes are never carried by the vehicle. At the top of the hill the vehicle is fitted with one of the shoes. At the bottom the shoe is left at the roadside to be picked up later by a cart going uphill. These wheeled skids are made by blacksmiths.

# 6 LOCK

According to Thomas Hennel,[10] wagons are named and classified according to the amount of turning which is possible in the fore-wheels, 'quarter-lock in which the movement of the wheels is limited by the straight sides of the waggon floor; half-lock in which a section is taken out of each side . . . three-quarter lock in which only the pole of the waggon impedes their movement; and full-lock waggons, in which the wheels are made small and can turn under the floor of the waggon.'

Although this classification gives us an idea of the side construction of a vehicle, it does not signify the amount of turning space required. Indeed, the classification often gives a false conception of the locking qualities of a wagon. A Dorset wagon, for example, with its straight frame can be described as a quarter-lock, yet it can turn in a space of 30 feet. A late nineteenth-century wagon from North-West Gloucestershire, on the other hand, with its waisted body can be called a half lock, yet it requires as much as 52 feet to turn.

The lock of a vehicle is determined by a number of factors, the prime factor being the side construction of the vehicle. Some wagons, like those of Dorset, are straight framed so that, on turning, the wheels soon rub against the frame. Others, like those of the Cotswolds, are waisted while some, like the old road wagons, have notched frames.

The second feature which determines lock is the position of the fore-wheels in relation to the body. If they are close to the body the lock is limited even though the vehicle may be waisted. An excellent example of this is the Monmouthshire wagon, where the angle of turn is no more than $17\frac{1}{2}$ degrees, although the vehicle is waisted. This is because the top of the wheels are no more than $4\frac{1}{2}$ inches from the body of the wagon. On the other hand, if the fore-wheels are a considerable distance from the body, the lock is far greater even though the body may be straight framed. The top of the Dorset wagon's fore-wheels are $8\frac{1}{2}$ inches from the body, and although the body is not waisted the angle of turn is 26 degrees.

The third factor affecting lock is the dish of the wheels. A wheel with a considerable dish can turn much more before rubbing against the body than a wheel which is perfectly cylindrical.

The last factor determining lock is the wheelbase of the vehicle, that is the distance between the centre of the fore-axle and the centre of the rear axle. Thus, the Dorset wagon with wheels turning at an angle of 26 degrees has a wheelbase of only 67 inches, and can turn in a space of 30 feet. A late example of an East Anglian wagon on the other hand has wheels able to turn at an angle of 35 degrees right under the body. Although with a larger wheelbase of 84 inches this wagon, too, can turn in 30 feet.

The following table gives the amount of space required for turning by each of twelve wagons which are now at the Museum of English Rural Life. The wheels of a wagon make four circles, the largest being that made by the outer front wheel. This dictates the minimum turning space required by the vehicle.[11]

| Origin of Wagon | Hennel Classification | Angle of turn (degrees) | Diameter of circle formed by outer front wheels at full lock (feet) |
|---|---|---|---|
| Suffolk . . . . | $\frac{3}{4}$ | 35 | 30 |
| Cambridge . . . | $\frac{3}{4}$ | 49 | 23 |
| Sussex . . . . | $\frac{1}{2}$ | 32 | 32 |
| Lincoln . . . . | $\frac{1}{2}$ | 31 | 31 |
| Wiltshire . . . | $\frac{1}{4}$ | 21 | 38 |
| Dorset . . . . | $\frac{1}{2}$ | 26 | 30 |
| Berkshire . . . | $\frac{1}{2}$ | $27\frac{1}{2}$ | 34 |
| Cotswold (South) . . | $\frac{1}{2}$ | $27\frac{1}{2}$ | 36 |
| Cotswold (North) . . | $\frac{1}{2}$ | 22 | 41 |
| Oxford . . . . | $\frac{1}{2}$ | 29 | 32 |
| North-West Gloucestershire | $\frac{1}{2}$ | $17\frac{1}{2}$ | 52 |
| Devon . . . . | $\frac{1}{4}$ | 27 | 33 |

# 7 LADDERS

There are three common ways of increasing the loading capacity of a vehicle. Firstly, the wagon may be equipped with sideboards to take the extra load at each side. Secondly, it may have vertical poles fitted to each corner of the wagon. Thirdly, there may be front and tail ladders which project beyond the body of the vehicle.

Sideboards have already been described, but one particular construction is especially interesting. On Lincolnshire, Nottinghamshire and some Herefordshire wagons the sideboards are removable and are only fitted to the wagons at harvest time. These frames consist of two stout poles about 10 feet long. At the front, two or more slats or keys join the side pieces together. At the bottom of the frame are a number of hooks, which fit into staples on the top rail of the wagon. Each wagon has two of these frames which meet at the centre and project forwards as sideboards and fore-ladders. This method of increasing the loading capacity is common on two-wheeled carts. In fact the whole idea of using such a device on four-wheeled wagons probably originated from the Scotch cart. The Scotch cart, a general-purpose farm vehicle, became popular in England in the late eighteenth and early nineteenth centuries. The load carrying capacity of its long body could be very greatly increased by fitting a wooden framework to the top. Presumably this method was later adapted to the box-wagons.

In the Vale of Gloucester and in South-East England straight poles are fitted at each corner of the wagon to hold the hay and corn sheaves in place as the vehicle moves. In Sussex, Kent and Surrey a wagon is equipped with four poles, each passing through a pair of staples on the outside of the wagon. At the front there are two pairs—one on the forebridge and the other on the front top-rail. At the back one pair is on the inner top-rail, the other on the rear cross-bar. In this way the poles, when fitted, are extensions to the wagon itself.

In the Vale of Gloucester the arrangement is rather different. The wagon may be equipped either with a fore-ladder or with a pair of poles at the front—they are rarely found at the back. In this case each pole passes through a pair of rings on the inside of the frontboard, one pair on the top-rail and the other on the floor of the wagon. In Hereford, poles may be fitted to the wagon as well as removable sideboards. The poles are not at each corner of the wagon but approximately 15 inches away from the front and back respectively. Again they pass through staples on the outer side of the body.

Fore- and tail ladders have been common on English farm vehicles since medieval times. The harvest cart in the Luttrell Psalter, for example had both types and the load overhangs both at the back and front.

These ladders are made of two strong side pieces joined by two or more thin slats or keys. The fore-ladder is fitted to the top of the sideboards either through a pair of staples or simply held under the spindles of a sideboard. This type of fore-ladder projects a foot or so over the horse's back, and when fitted is almost horizontal.

In other cases the ladder is fitted to the floor of the wagon and slants over the horse's back. This type of ladder is usually large, that of Wiltshire, for example, being as much as 6 feet 10 inches long.

Some ladders, notably those on East Anglian wagons, are held in place by a pair of brackets on the outside of the wagon body. East Anglian wagons have removable frontboards and a ladder may be used as an alternative to the top half or the whole of the frontboard.

Tail ladders are far less common than fore-ladders and many wagons, especially those of the Cotswolds and Oxfordshire, are rarely fitted with them. On others the tail ladder is much shorter than the front ladder and may be fitted either to the sideboards, the floor of the wagon, or to brackets on the outside in the same way as fore-ladders. Some may project backwards, carrying the load beyond the furthest stretch of the opened tailboard. The larger variety slant over the closed tailboard from attachments on the floor of the wagon.

# 8  PAINTING AND DECORATIONS

Some regions in Britain have adopted individual colours which are used not only for painting farm wagons but also for all other farm tools and implements. In the county of Rutland, for example, all farm equipment is painted ochre red, while in Cardiganshire a delicate shade of pink is universal. There are many other regions, however, where there is no traditional colour. Within these areas farm implements and wagons are painted in a variety of shades.

For example, although most Dorset wagons are painted yellow, blue wagons and blue-black wagons are often found within the county.

As a general rule, however, a large proportion of the wagons of a particular county or region are painted in one colour, the bodies of the vehicles being in one of the following three colours:

YELLOW. The wagons of the Cotswolds, Shropshire and mid-Wales are always painted in this colour, while most Dorset wagons have yellow bodies.

BLUE. This is by far the most popular colour. The shade varies from the dark, greenish blue of Lincolnshire wagons to the light blue of the Devon wagons. Other examples of blue wagons are those of Wiltshire, Sussex, Hereford and Suffolk.

BROWN. The wagons of Surrey, North-East Hampshire, Yorkshire and Hertfordshire are painted a dark brown while those of Huntingdonshire, Cambridgeshire, Rutland and East Leicestershire are painted a lighter shade of orange brown.

Besides these three main colours there are some vehicles, such as those of Kent and Norfolk, painted in a buff or stone colour.

Whatever the colour of the body, the undercarriages of all wagons are painted venetian red, often with the nave bonds picked out in black. The ironwork of the body and undercarriage is also picked out in black on many wagons.

In country workshops the paint is mixed by the craftsman himself. The pigments are ground on a stone slab with a series of rounded stone blocks known as mullers. The main constituent of the paint is white lead. The pigments used are ochre, venetian red and lamp black, together with a great variety of other colours depending on the taste of the craftsman and the tradition of his neighbourhood. In painting a new wagon in Sussex,[12] for example, the body is given two coats of lead colour followed by putty filling. Finally there is the top coat of prussian blue. The wheels are given two coats of red lead, while the undercarriage, the underpart and inside of the body are given two coats of venetian red, followed by putty filling and a top coat of venetian red. In addition, all the framework joints are coated with white lead prior to assembling. Sussex wagons were painted regularly every five or six years.

Many English farm wagons are decorated with painted lettering or chamfering, mainly on the frontboards and tailboards of the vehicles.

H

## (1) FRONTBOARD DESIGNS

The frontboards of wagons may be divided into the following broad categories:

### PLAIN UNDECORATED FRONTBOARDS

On some regional types the frontboards are perfectly plain and do not even bear the name of the owner. The Shropshire and Sussex wagons are examples of this type, where the frontboards have three upright ribs and a mid-rail—a construction similar to the panelled sides. Hertfordshire wagons, on the other hand, have shallow, spindled frontboards bearing no inscription of any kind.

### CHAMFERED FRONTBOARDS WITHOUT INSCRIPTION

In this case the frontboards are decorated with a series of carefully shaped notches and chamfers, but no inscription. The removable frontboard of the East Anglian wagon is kept in place by two cross-pieces pinned to the forward extension of the mid-rails and inner top-rails. These cross-pieces are no more than 5 inches wide and each is decorated with a series of regular chamfers on the lower edge. These chamfers are picked out in red or white, contrasting sharply with the blue of the remainder of the frontboard.

In Yorkshire the lower edge of the front top-rail is chamfered. The three vertical ribs and the forebridge are also chamfered. These notches are picked out in yellow or white and a complicated line design is painted on the top-rail.

### UNDECORATED LETTERING

On many wagons the name of the owner of the vehicle is inscribed on the frontboard, often accompanied by his full address and the date when the wagon was built.

The simplest form of undecorated lettering occurs on some Wiltshire wagons, where the shallow board is filled with the name of the owner.

In other parts of the country both the name and address of the owner appear on the frontboard. On the East Midlands wagon the centre of the board bears the inscription in small black letters, contrasting sharply with the ochre red of the remainder of the body. In the Cotswolds the inscription is in red or black on a yellow background, and the frontboard is so shallow that the name

(a) Sussex Wagon

(b) Hertfordshire Wagon

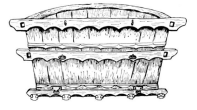

(c) East Anglian Wagon

(d) Yorkshire Wagon

(e) Somerset Wagon

(f) West Berkshire Wagon

(g) Lincolnshire Wagon

(h) Rutland Wagon

(i) Vale of Berkeley Wagon

(j) Huntingdonshire Barge Wagon

25  Decorations on frontboards

and address of the owner completely fills it. In Somerset, and sometimes in Devon, the inscription is painted on a small board attached to the two upright ribs of the frontboard. At the bottom the name of the builder and date of construction appears in small lettering. The supplementary board, itself of the same depth as the frontboard, is painted in a contrasting colour to the rest of the body. The frontboards of the later plank-sided Lincolnshire wagons which are inscribed but not decorated, may be regarded as degenerate examples of the elaborately decorated and inscribed fronts of the older spindle-sided wagons. Generally, the boards carry the name and address of the owner, the date when the wagon was built, the name of the maker, usually on an iron trade-plate, and the name of the painter. In Staffordshire and North Warwickshire it is customary to fit two curving pieces of iron running from the front top-rail to the forebridge. Inside the iron strips the frontboard is painted red, while outside it is blue with the name and address of the owner in black paint.

## DECORATED LETTERING

Within Great Britain a variety of elaborately decorated and lettered frontboards may be found, but in this section only a few can be mentioned. Although within a region great variation may occur in detailed design, a general style of decoration is recognisable.

On many of the older wagons in the East Midlands the name and address of the owner appears on an oval or circular plate, painted in the centre of the board. This is in black, with the inscription elaborately painted in yellow or white with red shading. Below the painted oval there is the name and address of the builder, together with the date of the construction of the wagon. The front ends of the two mid-rails are joined to the frontboard by two pairs of moulded sheet-metal fleur-de-lis tie bolts. These metal pieces are picked out in black paint. On some wagons there is a great deal of line decoration interwoven with the lettering.

In the Lower Severn Valley and South Wales the frontboard decoration is somewhat different. There the inscription is painted on an elaborately decorated semi-circular board, nailed to the lower half of the frontboard. This red painted wooden plate bears the name and address of the owner as well as the date of construction.

Apart from wagons which bear some form of geometrical design on the frontboards, there are others bearing complicated leaf designs painted in a variety of contrasting colours. In Dorset, for example, wagon frontboards are highly decorated with elaborate lettering and leaf patterns in black, red and white, contrasting sharply with the yellow of the body. In Lincolnshire great pride was taken in lettering the frontboards and elaborate leaf designs encircled the name and address of the owner of the vehicle, the date of construction and the name of the builder.

A form of frontboard design introduced into Eastern England in the nineteenth century was the spectacle. Tebbutt[13] notes that this design in its simplest form 'was originally made of two layers of boarding, the outside layer having cut out of it the shape of spectacles, and the back layer thus exposed being painted a contrasting colour'. In time this method of carved decoration degenerated into the simple painting of the spectacle design on a single layer boarded front. This design was introduced to England from Scotland on the well-known Scotch cart when large numbers of these were imported into the eastern counties during the early and mid-nineteenth century. Local wheelwrights took over this design and used it on the frontboards of their carts and wagons. However, the adoption of this design must have been gradual, for as far as wagons are concerned, it occurs only on plank-sided wagons built during the last quarter of the nineteenth century in East Anglia and the East Midlands. In eastern England various degenerate forms of spectacle design are found on a large number of wagons. It varies in shape from elongated oval strips to a pair of unconnected circles.[14]

LETTERED BOARD ATTACHED

The wagons of Hereford and those of similar types from the surrounding counties are generally equipped with plain panelled frontboards. The name and address of the owner is inscribed in black paint on a white board, attached to the top of the frontboard. The edges of the board are tacked on to the ribs, and the lower edge considerably notched, the notches being picked out in black paint.

## (2) TAILBOARD DESIGNS

The tailboards of English wagons are not generally decorated as much as the frontboards. On the majority of wagons the tailboards are either spindled or planked and are not decorated. In all cases the board is removable, and is held in place by eye bolts fitting into iron projections on the rear cross-bar, or by the iron support fitting into a pair of staples on the rear cross-bar.

In some places, notably East Anglia, tailboards are rare but, when used, they are basically of the same design as the frontboards. Each is kept in place by one or two horizontal cross-pieces which are notched and painted in a contrasting colour on the lower edge.

Occasionally some indication of the owner's name is found on tailboards, but rarely does it give more than his initials. For example, on a Cotswold wagon at the Museum of English Rural Life, the name and address of the owner, 'Robert Howard, Eastleach, Gloucestershire', is found on the frontboard, but only the initials 'R.H.' appear on the tailboard.

Only on Dorset wagons is there elaborate lettering and design on the tailboard. The wagons of that county have the name and address of the painter and builder of the wagon as well as the date when the wagon was last painted. These are, more often than not, accompanied by extremely intricate painted scroll work in a variety of colours. The boards are yellow, while the lettering and date are usually black, shaded with blue or red. The scroll decorations are in red shaded with black.

## (3) SIDE DECORATIONS

Rarely does one find elaborate decoration on the sides of a wagon although, in some cases, there is a certain amount of chamfering and carving. On East Anglian wagons, for example, a series of straight parallel lines are cut in the main frame running the whole length of the wagon. The lines are repeated on the mid-rail, the top rails and on the sharply sloping side-boards.

The later plank-sided Hereford wagons have two deep grooves running the whole length of the body. These are painted in red or white to contrast with the blue of the side planks. They do not follow the junction of the three side planks as would be expected but follow the line of the top rail, cutting across the junction of the planks. These grooves are purely decorative and probably represent the mid-rails of the older panel-sided vehicles.

The outer top-rails of some wagons, such as those of Worcestershire, are notched along the whole length of the wagon. The South Midlands bow wagon has a highly ornamented body. Its outstanding characteristic is the side frame, chamfered to cut down the weight of a wagon used in a hilly region. Where the rear side plank ends just above the waist of the vehicle it is delicately carved along its forward edge with the small notches picked out in red paint. The front of the side plank is also notched and painted.

In many districts the name and address of the owner is written on the sides of the wagons, either directly on the side plank as in South Midland wagons, or on a wooden board nailed to the side planks at the front. In most cases this plate is rectangular in shape, and may be painted the same colour as the remainder of the wagon, as on Lincolnshire wagons, or black, as on Sussex wagons. In both cases the lettering is in white or yellow. In Hereford and the surrounding counties, however, the name-plate is a strip of wood coloured white, with the name and address of the owner written in black paint. This wooden plate which follows the curve of the wagon is nailed to the side just below the inner top-rail.

## NOTES

1. [With the construction of larger vehicles during the sixteenth and seventeenth centuries stronger naves were used to accommodate the thick wooden arms of the axles and to bear the pressure of an increased number of spokes. Barrel-shaped naves became common and continued in use until iron axle arms were fitted and spoke mortices were staggered. The nave was commonly 15 inches in diameter and 14 or 15 inches long, holding an axle arm of 5 or 6 inches thick. Ten, twelve or even fourteen slanting spoke mortices were cut in the face of the nave. The mortices tapered from 2 inches at the front to $1\frac{1}{4}$ inches at the rear and were cut to a depth of $3\frac{1}{2}$ or $3\frac{3}{4}$ inches. The main object of tapering the mortice was to avoid weakening the nave. With iron axle arms and decreased dish a cylindrical nave came into use. Although the cylindrical nave is simpler to make than the larger, barrel-shaped hub, the methods of construction and the tools used are basically the same.]
2. EDLIN, H. L.   *Woodland Crafts in Britain*–London 1949–p. 53 *et seq.*
3. MERCER, H. C.   *Ancient Carpenters Tools*–Doylestown, Pa. 1951–pp. 222-6
4. STURT, G.   *The Wheelwright's Shop*–Cambridge 1923–p. 137–'To secure foreway, the arms were given a bedding, a slight angle forwards, so that the wheels on them had a tendency to draw themselves inwards at every turn. It was not much. There was no rule or scale of foreway known in my shop . . . And so it was that every wheel on every wagon or cart had another motion besides going round and round. In addition to that, the wheel was always sliding to and fro on its wide greased axle arm—now driven "click" against the lynch pin; and next second trying once more (thanks to foreway) to pass the shoulder—"clack" on the opposite end of the arm, only to be sent back, "click" to the lynch-pin again.'
5. *Ibid.*–pp. 135-6
6. [G. W. Weller, a wheelwright, in a letter to R. A. Salaman in 1952 gives the following account of setting out a wooden axle,
   'Example shows a pair of $5\frac{1}{2}$ inches by $3\frac{1}{2}$ inches set out to carry a pair of wheels 4 feet 10 inches high with 2 inches dish. Effective length of

spoke 18 inches measured from nave to felloe. Nave of wheels 15 inches diameter and 15 inches long.

Half diameter of wheel 2 feet 5 inches less half diameter of nave 7½ inches less depth of felloe = 3½ inches. This gives length of spoke between hubs and junction with felloes.

The wheels have to be 4 feet between shoulders, or hind ends of naves. In practice the diagram is set out full size on a setting out board measuring 6 feet long and 7 inches wide.

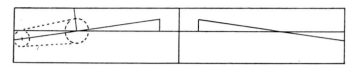

1. The distance between the shoulders is set on the straight line (4 feet).

2. The length of spoke is transferred inwards from the shoulders and marked.

3. The dish (2 inches) is set square up from the inner end of the spoke length.

4. The length of box is measured from the shoulder to linch pin.

A line is struck from the dish measure, through the shoulder mark on straight line, to linch pin length.

The diameter of the arms (5½ inches and 3½ inches) is marked on the inclined lines at points shown. A line drawn at the bottom of these diameters shows inclination of the axle arms.

A gauge or set is made to the correct inclination of the arms. This gauge transferred to the underside of the work enables the workman to keep his work true.']

7. STURT, G. *op. cit.*–p. 33
8. *Ibid.*–p. 63
9. STEVENSON, W. *General View of the Agriculture . . . of Dorset*–London 1815–p. 165
10. HENNEL, T. *Change in the Farm*–Cambridge 1934–p. 28
11. The radius of the turning circles described by each wheel of a wagon may be calculated from the following formulae,

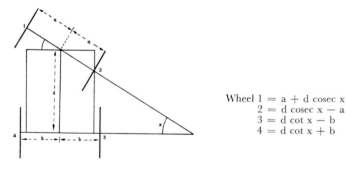

Wheel 1 = a + d cosec x
       2 = d cosec x − a
       3 = d cot x − b
       4 = d cot x + b

The area of the circle described by wheel 1 when the front wheels are on full lock is the minimum ground area required for turning the wagon through a complete circle.

12. H. E. Philpot, private communication
13. TEBBUTT, C. F. 'Some Cart and Wagon Decorations of the British Isles and Eire', *Man*–August 1955–Vol. LV, pp. 113-7

14. EVANS, E. E. *Irish Folk Ways*–London 1957–p. 178–'C. F. Tebbutt is inclined to seek an origin in the magic *oculi* such as occur in prehistoric pottery and to this day on Mediterranean boats, but it is difficult to see the connecting links. If one looks for a mere practical origin it could be argued that when freshly painted front-boards were first fitted to carts they were marked and damaged by the rump and tail of the horse. At any rate I have occasionally noticed a spectacle-like pattern rubbed by the horse on to the paint of Scots carts not fitted with the recessed front-board. On this theory the board was strengthened by means of panelling while at the same time the shaped recess gave the horse's rear a little more space and saved the paint.'

# BOX WAGONS

## V

### 1  EASTERN COUNTIES WAGONS

The largest type of English farm wagon is found in the eastern counties, in a region that includes East Anglia, Lincolnshire, Nottinghamshire and parts of Derbyshire. The main characteristics of the wagon are:

> The large, narrow tyred wheels which make the floor of the wagon high and rather difficult to load.
>
> The deep body with one or two mid-rails.
>
> The complete absence of sideboards in the Northern (Lincoln) variety and the narrow single-railed sideboards of the southern (East Anglian) variety.
>
> The curving top-rails which rise steeply to a high frontboard and less steeply to the tailboard, giving the wagon a curved profile.
>
> The waisted frame and the notching of the lower side planks below the lower mid-rail to provide greater lock.

On the basis of differences in body construction the wagons of the eastern counties can be divided into two distinct types, (1) The spindle-sided northern or Lincolnshire wagon, (2) The panel-sided southern or East Anglian wagon.

### (1)  LINCOLNSHIRE WAGON

The large spindle-sided Lincolnshire wagon is used in the following counties:

> LINCOLNSHIRE. It is used throughout the Kesteven and Lindsey sections of the county. In the Holland division, where small farms predominate, wagons are rare, but those that do occur are of the Lincolnshire type.
>
> RUTLAND. The Lincolnshire wagon extends into Rutland, where it is found alongside the smaller East Midlands type on the larger farms of the county.
>
> LEICESTERSHIRE. The Lincolnshire wagon is limited to the Vale of Belvoir.
>
> NOTTINGHAMSHIRE. It is used throughout the county.

DERBYSHIRE. The Lincolnshire wagon is rare in the county, and occurs only in the valleys of the Peak District.

YORKSHIRE. A few wagons of Lincolnshire type were seen in Holderness, although it could not be ascertained whether these were made in the district.

Throughout this large region, stretching from the Fens in the south to the Humber and industrial South Yorkshire in the north, the wagons are remarkably uniform in size and construction. Although, on the whole, the Lincolnshire wagon is well suited to the topography of much of the area, its high, narrow build makes it less suited to the abrupt changes of slope in the Wolds and Lincoln Edge. One farmer from the Horncastle district mentioned that on one occasion his two wagons capsized twelve times during the course of one day's harvesting. Yet the vehicles used in the Wolds are the same in every detail as those used in the flatter parts of the eastern counties, and local craftsmen have made no attempt to build lower and wider wagons more suited to the hilly districts.

In construction the Lincolnshire wagon is the simplest of all English box wagons, and closely resembles the larger farm wagon

26　Lincolnshire Wagon from Leadenham

of the Low Countries. Like its continental counterpart it has a deep, spindle-sided body, and a tall, highly-decorated frontboard. It has no sideboards to take the overhanging load, while the large wheels are narrow tyred. Vehicles of continental origin were used throughout England as road wagons from the sixteenth century onwards. When four-wheeled vehicles were first used for agricultural transport, the various regions of Britain adapted the existing road wagons to the nature of their own countryside. In Lincolnshire and the adjoining counties no such adaptation occurred, and the farm wagons continued in much the same tradition as the older road wagons.

### WHEELS AND AXLES

The wheels of a Lincolnshire wagon are large. Both on the marshy soils of the coast and on the harder soils of the Wolds, narrow-tyred, three-inch wheels are always found. Each wheel is tyred with a single hoop and during this survey not a single straked wheel was seen anywhere in the region.

Although hoop tyring, probably the most advanced method of tyring the wheels of farm vehicles, made an early appearance, other obsolete features of construction persisted until a late date. In some parts of the country iron axles had already replaced wooden axles by the 1850s, but in Lincolnshire all-wooden axles persisted for as long as village craftsmen built wagons. A plank-sided wagon seen at Burwell in Lindsey was built as recently as 1935, but was still equipped with thick wooden axles.

Because of the persistence of wooden axles, massive barrel-shaped hubs were needed, often with a diameter of 16 inches. Each hub is equipped with two iron bonds. A narrow hind-bond and an unusually wide breast-bond are fitted just in front of the spoke mortices. Cylindrical hubs of small diameter with staggered spoke mortices never made an appearance in Lincolnshire.

### UNDERCARRIAGE

The fore-wheels of a large diameter are set well forward to increase the lock of the wagon. For this reason the simply constructed forecarriage is very short. On a wagon seen at Sleaford the whole forecarriage measured no more than 50 inches from back to front, and 35 inches of this was behind the fore-axle. The forecarriage itself consists of two outside hounds joined together by a slightly curved slider bar, but not connected by any form of shutter. A pair of locking chains join the hounds to the side frames of the vehicle and prevent the wheels from rubbing against the body.

The shafts are attached to the forecarriage by a 1 inch iron shaft pin which passes through the ends of the hounds and shafts. Despite the size and weight of the wagon, the forecarriage never ends in a splinter bar and consequently double shafts are never found. When two or more horses are required to draw the wagon they are harnessed in tandem, never in line abreast.

The pole joining the forecarriage to the rear carriage is straight and ends just behind the rear axle. This pole is no more than $3\frac{1}{2}$ inches square and seems extremely light considering the heavy body it has to support. Neither its lower surface nor the top surface of the slider bar are covered with iron plating, and there is a very real danger of some part of the undercarriage breaking if the wagon is in constant use. During this survey at least ten wagons were seen with broken coupling poles. It seems surprising that so much attention should be paid to the design and construction of the body while that of the undercarriage is largely neglected.

One great disadvantage of all Eastern Counties wagons is that the floor level is too high for easy loading. In an attempt to lower the floor level, the bolster and pillows are made shallow. Each member is no more than 3 inches deep and 8 inches wide, consequently the top rail of the wagon is only about 6 feet 6 inches above ground level. In sharp contrast to the inadequate undercarriage the body of the Lincolnshire wagon is large, capacious and heavy.

FRAMEWORK

The framework of the body is heavily constructed, the main cross-bar being a straight piece of timber 5 inches wide and 4 inches thick. Into this the curving, forward side frames and the rear summers are morticed.

The forward side frames, made from naturally curving oak $5\frac{1}{2}$ inches wide and 4 inches deep, are morticed into the forebridge, iron plates being hammered over each mortice. The surfaces curve inwards and pass through the centre of the main cross-bar, ending just behind it.

The rear side frames, made of 4 inch square oak, are straight, and pass over the main cross-bar to end in a chamfered edge just in front of the bar. A $\frac{1}{2}$ inch coach bolt is driven through the rear side frame and cross-bar.

In addition to the heavy side frames which run to the waist, a wide central summer runs from the forebridge through the main cross-bar to the rear cross-bar. This summer is 5 inches wide at the front. It widens gradually to 6 inches at the king pin and then narrows to 5 inches at the main cross-bar. Two inside summers on either side of the central summer run from the rear to the main cross-bar. Since the Lincolnshire wagon is used for many purposes besides harvesting hay and corn, it generally has a long boarded floor resting on oak keys morticed at right angles to the frames and summers.

## BODY

The sides are built with two side planks separated by one mid-rail and capped by a curving top-rail. The lower plank is notched to take the locking fore-wheels, the inner side of the waist being covered by a triangular construction fitted both to the side planks and to the floor on the inside of the wagon. The main feature which distinguishes the Lincolnshire wagon from its close relative, the East Anglian wagon, is the large number of spindles morticed into the rails and nailed to the side plank. These are long and slope sharply at the front, become more upright towards the centre cross-bar, and thence slope backwards more and more to the rear of the vehicle. In south-eastern Lincolnshire, towards the Leicestershire-Rutland border, the wagons, which are generally the same shape as those used in the remainder of Lincolnshire, have two mid-rails. These may be regarded as transitional types between the curved, single mid-railed, true Lincolnshire wagons, and the straight-sided double mid-railed East Midlands wagons.

As a general rule, the Lincolnshire wagon has only two pairs of iron side-supports of square cross-section. Delicate and elaborate iron scroll-work is by no means uncommon on the side-supports of the older wagons. The first pair of supports are found at the main cross-bar, joining the frame to the top-rail, while the second pair is found on the rear cross-bar. Unlike all other regional types, Lincolnshire wagons are not fitted with overhanging sideboards, a feature that shows unmistakably their close relationship to continental farm wagons. In order to increase the load carrying capacity of the vehicle a set of harvest frames can be fitted.

Undoubtedly this idea was copied from the Scotch *coup cart*, which became common in the county during the early nineteenth century. The frames are made in two sections. The front section, which covers the front half of the wagon, fits into two pairs of staples on the top-rail. The rear section runs from the centre of the wagon to beyond the tailboard, and is also held in place by two pairs of staples on the top-rail. The harvest frame, which projects well beyond the sides of the wagon, combines the advantages of sideboards and end ladders, and the load carrying capacity of the vehicle is greatly increased. For example, in Nottinghamshire the loading space of a wagon in the early years of the nineteenth century was only 13 feet long and 5 feet wide. By adding harvest frames the load-carrying platform was increased to 24 feet long and 8 feet wide.[1] When not required for harvesting the frames are, of course, removed.

## COLOURS

The wagons are well painted, prussian blue being the most common colour for the body, except in the west where a brick-red colour predominates. The tall, fixed frontboard is highly decorated with scroll-work. The name and address of the owner, the date of building, the name of the builder, and even the date of the last painting are carefully interwoven in a contrasting light colour.

The lofty frontboard, the sharply curving top-rails and the great care taken in painting make the Lincolnshire wagon the most elegant of all English box wagons.

## (2) EAST ANGLIAN WAGON

The panel-sided East Anglian wagon, the largest of all English farm vehicles, is used in the following counties:

ESSEX. The East Anglian wagon is found throughout the northern part of the county, that is in the arable area on the glacial drift soils. In South Essex, where the glacial drift soils give place to the clays and loams of the London Basin, and arable farming gives place to dairying and market gardening, wagons are rare. The few that do occur in South Essex are of East Anglian type.

SUFFOLK. The East Anglian wagon is found throughout the county.

NORFOLK. The East Anglian wagon is used throughout the

county, but in North Norfolk the vehicles are smaller and lighter than in the other counties, although similar in shape and construction.

CAMBRIDGESHIRE. The East Anglian wagon is found in the eastern portion of the county, but in the Fens wagons are rare.

HERTFORDSHIRE. In the south-west no distinct line can be drawn between the panel-sided East Anglian wagon and the spindle-sided East Midlands wagon. Both types were seen in the Bishops Stortford area of Hertfordshire but this district probably marks the western limit of the East Anglian wagon.

Since East Anglia is predominantly an arable region of low relief, very large wagons are used in great numbers throughout the region. At harvest time the need for vehicles to carry the heavy corn crops is so great that two-wheeled dung carts are converted into wagons by the addition of pairs of fore-wheels and forecarriages. These vehicles 'partaking of both a cart and wagon' are known as 'Hermaphrodites' (Fig. 28), and they are used throughout the eastern counties of England from North Essex to Nottinghamshire.[2]

In general appearance the East Anglian wagon is massive, with wheels of large diameter and a very deep body. The top-rails curve

27   East Anglian Wagon from Sible Headingham, Essex

gently from the frontboard to the centre of the wagon and then rise very gradually to a deep tailboard. The curved profile of the East Anglian wagon is less noticeable than that of the Lincolnshire wagon but more so than the East Midlands type. The wagon creates a general impression of massive solidity compared with the delicate lines of the Lincolnshire vehicles.

## WHEELS AND AXLES

The wheels of an East Anglian wagon are very large and a diameter of 6 feet is quite common for rear wheels. Even in the late nineteenth century when fore-wheels were made small enough to lock right under the body, the rear wheels still retained their great diameter. The wheels are dished, but not as much as the wheels of wagons designed for use in hilly districts such as the Cotswolds. More often than not the wheels of pre-1880 wagons are straked, each strake being held by four or five square-headed nails at either end. Since there is little really heavy land in the region, double-tyred broad wheels are unknown and a tyre width of 3 or $3\frac{1}{2}$ inches is the most common.

An unusual feature of many East Anglian wagons is that the track of the fore-wheels is often much less than the track of the rear wheels. There is no structural reason for this, and the evidence points to it being a long established custom. In the eighteenth century wagons that rolled a broad track were subject to lighter tolls than narrow-wheeled vehicles. After 1774, when an Act of Parliament gave complete freedom from tolls to all wagons rolling a track of more than 16 inches, vehicles were built with shorter fore-axles so that a 9 inch tyre on each wheel would roll a track of at least 16 inches. Although nineteenth-century farm wagons in East Anglia were equipped with narrow wheels, for no apparent reason, the custom of using short fore-axles continued until the end of the century.

The wheel naves are large, each being 15 inches long and 16 inches in diameter. These large naves persisted because all-wooden axles were used on all farm wagons until the appearance of standardised barge wagons in the last decade of the nineteenth century.[3] The axle arms have a slight downward pitch which varies according to the dish of the wheels. The forward cant of the arms is not very noticeable.

I

28   Hermaphrodite from Gillingham, Norfolk

## UNDERCARRIAGE

The forecarriage consists of two straight outside hounds joining the straight splinter bar to the slightly curved, iron-lined slider bar. Each hound is bolted to the underside of the bar by a ½ inch coach bolt. Two inside hounds run from the splinter bar to just behind the axle. Two sets of shutters are morticed into the hounds so that the whole forecarriage is very strongly constructed. The carriage itself is set well forward under the body of the wagon so that the leading fore-wheel easily clears the front corner of the vehicle when locking.

The coupling pole, a thick block of timber some 5 inches square, curves over the iron-plated slider and curves backwards beyond the rear axle to be loosely pinned to the rear cross-bar of the body framing.

The bolster laid over the fore-axle and the pillow bolted to the wagon frame are both massive, about 14 inches wide and 6 inches deep, and are cut from a heavy slab of beech or elm. Both the bolster and pillow are carved to an unusual shape, having a semi-circular bulge on the rear face around the king pin. This arrangement steadies the body of the wagon in turning and gives additional support to the king pin.

Since East Anglian wagons are extremely heavy, two pairs of shafts are fitted. Each shaft is attached to an unusually wide splinter bar by two shaft pins which pass through two sets of barrel eyes fitted on to the front of the bar.

## FRAMEWORK

The body of the East-Anglian wagon is capacious and extremely well constructed. The centre cross-bar is a heavy, curving timber, 6 inches wide and 10 inches deep, plated with iron along its top surface. Morticed into it are the main rear side frames, which continue in a straight line to the rear cross-bar. In addition, three inside summers, slightly lighter than the side frames, are morticed into each member. The forebridge is a gently curving piece of timber joined to the main cross-bar by two heavy front side frames. Each side frame is made from a heavy timber curved inwards to provide a waist, and morticed near to the centre of the main cross-bar. A heavy inside summer extends from the centre of the forebridge to the centre of the cross-bar and, like the pillow and the bolster, it is bulged around the king pin. In addition, two lighter inside summers run from the forebridge to the curving rear section of the front side frames, being morticed to each member. (Fig. 27.) Thus the framework is constructed from extremely heavy pieces of timber, and is probably the strongest on any waisted wagon. The great weight of the vehicle is no real disadvantage in a countryside which is flat enough to make traction relatively easy. Large wheels, of course, do contribute to ease of traction, but on the whole the East Anglian wagon would be quite unsuitable for many other parts of the country because of its weight and size.

The floorboards of all East Anglian wagons are fitted cross-ways, perhaps reducing the weight of the vehicle. Extra pieces of timber for long boarding are unnecessary, since the East Anglian wagon is used almost exclusively for harvesting. In areas where wagons are used for carrying gravel, sand, etc., long-boarding is needed to provide a surface for shovelling the load from the wagon, but when wagons are used for harvesting a smooth shovelling surface is unnecessary. For everything but harvesting two-wheeled tumbrils are used and invariably they have long-boarded floors.

### BODY

The panel-sided body has one mid-rail which is square in cross-section at the centre but circular at the front and back where it projects a few inches beyond the planking of the sides. The upright standards joining the inner top-rail to the mid-rail and main frame are delicately chamfered to give the body of the wagon a panelled appearance. At the front the standards slope forward, but become gradually more upright towards the rear of the vehicle. Each side of the wagon has eight or nine standards nailed to the side planks and joining the frame, mid-rail and inner top-rail. Each standard is morticed into the various members. The main side supports of the wagon are made of iron and resemble an inverted Y, the first being fitted to the centre cross-bar and inner top-rail, the second to the rear cross-bar and inner top-rail. The four or five wooden supports are of two types. The first type is a combined side and sideboard support running the whole depth of the wagon. It is kept in place by a round metal bracket on the main frame and is riveted or bolted to the mid-rail and inner top-rail. The second type of side support runs from the mid-rail to the inner top-rail and acts mainly as a support for the sideboard. The two types alternate along the length of the wagon.

The most common method of side construction is the waisted framework combined with a notched body. The lower plank of the wagon side is notched so that the wheels can lock into the waist but in most cases the notch does not extend upwards as far as the mid-rail. Wagons with small fore-wheels may have no notch at all on the lower plank, even though they have a waisted framework.

The sideboards are very narrow and slope steeply from the outer to the inner top-rail. The span between these rails is filled with elm planks to produce solid sideboards.

The frontboard is elaborately decorated. Chamfered designs are more usual than an inscription and the name of the owner is rarely found on the frontboard although a small painted board is common on the side of the wagon. A very unusual feature not found on any other regional type of wagon is that both the frontboard and tailboard are removable. When the end boards are removed, they may be replaced by two bars, one connecting the mid-rails, the other connecting the inner top-rails.

As may be seen from Fig. 25 the frontboard is elaborately shaped. On the other hand the tailboard is much plainer, consisting of a number of inter-slotted slats kept together by one or more chamfered horizontal pieces of timber.

COLOURS

The East Anglian wagon is the most highly decorated of all English farm vehicles. In the greater part of the region the wagons are painted blue, but in northern Norfolk where a smaller wagon predominates the vehicles are painted a stone colour.

## BRAKING

Each wagon is equipped with a chain fitted to the rear side frame. This is passed round one of the rear wheels for travelling downhill, while a dog stick fitted to the rear axle is used to prevent the vehicle from slipping backwards on an uphill journey.

# 2 EAST MIDLANDS WAGONS

From Hertfordshire to North Leicestershire the same type of wagon occurs with remarkable uniformity. The main characteristics are:

Large wheels, with a tyre of no more than 5 inches.

A deep, spindle-sided body with two mid-rails.

In profile the top of the wagon is almost horizontal and the end boards are of the same depth as the rest of the wagon body.

The main side supports are wooden and a number of curved iron brackets join the upper mid-rail to the outer top-rail.

The side-frames are either straight, very slightly waisted or slightly notched.

The wide splinter bar with barrel eyes to accommodate one or two pairs of shafts.

On the basis of size the wagons of the East Midlands may be divided into two types, (1) the southern variety or Hertfordshire wagon, (2) The northern variety or Rutland wagon.

## (1) HERTFORDSHIRE WAGON

The Hertfordshire wagon is a small, spindle-sided, double mid-railed wagon used throughout rural Hertfordshire, South Bedfordshire and East Buckinghamshire. Since much of Hertfordshire is urban, wagons are rare, especially in the south of the county where market gardening and dairying play an important part in the economy. The two-wheeled Scotch cart replaced the traditional harvest wagon at an early date and only a few examples of traditional vehicles still survive.

As far as can be ascertained, the only wagons used in the south of the county during the last hundred years were those used to carry hay, corn, vegetables and other materials to the London markets. These may still be seen on the farms of Hertfordshire and Middlesex, and they were based undoubtedly on the design of the old farm wagons used throughout this region. These vehicles, heavily loaded with farm produce, would leave for the London markets early in the morning and return to Hertfordshire later on the same day. As the status and prosperity of the farmer was shown by the condition of his road wagon, the majority of these vehicles were well painted. A number of harvest wagons, however, were seen in the north and west of the county, and it is reasonable to assume that at one time vehicles of this type were used in South

29   Hertfordshire Wagon from Harpenden

Hertfordshire. These harvest wagons are like the ubiquitous market wagons in many ways and are also similar in shape and construction to the wagons of Surrey and other counties south of the Thames. Indeed this similarity between the Hertfordshire and Surrey wagon suggests that spindle-sided double mid-railed wagons were used not only in South Hertfordshire but in Middlesex as well.

With two exceptions all the Hertfordshire wagons seen and measured during the course of this survey came from the well dissected, overwhelmingly arable East Hertfordshire Plain. In the south-west, around Tring, the Hertfordshire wagon gives place to the South Midlands bow wagon. Farther north it penetrates into Buckinghamshire as far westward as the Missenden district, where it is again replaced by bow wagons. In the north there is no clear distinction between the Hertford and Rutland wagons, but northwards through Bedfordshire the vehicles become successively larger and more elaborately decorated.

In general appearance the Hertfordshire wagon is short and stubby, while in construction it is light and straight lined.

## WHEELS AND AXLES

Since most of the wagons are found on relatively light, arable soils, broad wheels are unknown and a tyre width of $2\frac{1}{2}$ or 3 inches seems to be common everywhere. Because the area lies on the main routes from London, it is agriculturally progressive so that antiquated features such as solid wooden axles and massive hubs had disappeared by the 1850s. All the fifteen wagons seen during this survey were equipped with iron axles. The practice of staggering spoke mortices and building cylindrical hubs appeared in the region by the early 1860s, anticipating the use of this technique in other parts of the country by many years. The small naves, rarely more than 12 inches in diameter, the shallow felloes and narrow tyres give the wheels of the Hertfordshire wagon an almost coach-like appearance in sharp contrast to the heavy wheels of the Rutland wagon.

## UNDERCARRIAGE

The forecarriage consists of two hounds morticed into a straight splinter bar running to a slightly curved iron-plated slider bar, where a pair of $\frac{1}{2}$ inch coach bolts join the various members together. The forecarriage ends in a splinter bar even though the wagons are small and light, requiring only one pair of shafts.

The pole joining the forecarriage to the rear carriage is short and straight and continues beyond the line of the rear axle and is pinned on the rear cross-bar.

Both the bolster and pillows are shallow, each being no more than 3 inches deep and 4 inches wide. Because of its straight frame and slightly dished wheels the wagon suffers from very poor lock and the wheels soon rub against the frame when turning. On some vehicles an attempt has been made to increase the lock by raising the floor level of the wagon. This is done either by fitting a pair

of wooden blocks to the main side-frames between the body and
the front pillow, or by putting a second bolster between the axle
and the forecarriage. This second device is particularly popular in
the Bishops Stortford district of East Hertfordshire. By increasing
the depth of the undercarriage in this way the floor of the wagon is
raised without a reduction in wheel size.

## FRAMEWORK

In Hertfordshire the wagon is used exclusively for harvesting, the
Scotch cart having replaced it for all other farm cartage in the
mid-nineteenth century. For this reason all the wagons are
equipped with cross-boarded floors. The body framework consists
of two 3 inch square timbers running in a straight line from the
forebridge to the rear cross-bar. In addition, running parallel to the
side frames, are two summers, each 2 inches square, which again
join the forebridge to the rear cross-bar.

On some wagons slightly heavier timbers are used for the side
frames. When the fore-wheels are designed to rub against the
body, the frames are notched in both a vertical and horizontal
plane to increase the lock. The slight vertical notch is achieved by
using naturally curving wood, while the horizontal notch is made
by cutting a shallow depression in the frame. Since the notching
leads to weakness in the side-frame, a heavy block of wood 3 inches
wide and 18 inches long is bolted to the frame on the inner side of
the notch.

The light framework of the wagon is also strengthened in some
cases—especially on vehicles after 1870—by bolting a second frame
underneath the main frame. This is rarely more than 2 inches deep
and has both the pillows and main cross-bar bolted to it.

## BODY

The body of the wagon is short but deep, and is constructed of
three narrow side planks separated by two mid-rails. Each mid-rail
is straight, measures 2 inches square and is bolted to a corresponding
cross-piece on the frontboard.

A large number of wooden, or even iron, spindles run from the
side frames through the two mid-rails to the inner top-rail of the
wagon. Both top-rails are constructed from 2 inch square timber,
while the sideboards consist of two other rails running the whole
length of the wagon parallel to the top-rails.

There are five wooden side supports on either side of the vehicle. The main support is morticed to the cross-bar, curves towards the two mid-rails and is held in place on the inner top-rail by a bolt. The rear support, morticed to the rear cross-bar, runs directly to the upper mid-rail and thence to the top-rail. The three intermediate supports join the frame to the three rails and are bolted to each, the bottom being held in a half round socket firmly embedded between the frame and side planks. Plainly forged sideboard supports with hooks for tying wagon lines run to the outer top-rail from the junction of the side supports and the upper mid-rail.

The Hertfordshire wagon is distinguished from its close relatives, the Surrey and Rutland wagons, by the frontboard which covers only the section below the upper mid-rail and never covers the whole depth of the body. This board is plain and undecorated and, like the wagon sides, is built up with round spindles of iron or wood. The wagon is equipped with a fore-ladder which fits into a pair of curved iron brackets bolted near the front of the upper mid-rail and frame. The ladder itself rests on the front top-rail which is thick and heavy to take the weight.

Tailboards are uncommon, but when fitted are similar in construction to the frontboards, except that they cover the whole depth of the body. The tailboard is attached to the wagon by a pair of pins fitting into barrel eyes on the cross-bar, so that its exact position can be adjusted according to the user's requirements. Although tailboards are relatively rare, most wagons are fitted with rear ladders. The ladder is quite short, its ends fitting into a pair of hooks on the main frame. It rests on the rear cross-bar and being nearly horizontal has the effect of lengthening the floor of the wagon.

COLOURS

The Hertfordshire wagon is painted dark brown, with buff wheels and undercarriage. All the line decorations on the body, especially on the front top-rail, are in buff or yellow. Road wagons, on the other hand, are painted in a great variety of colours, blue being the most popular.

BRAKING

The wagon is equipped with a chain for locking one of the rear wheels and also with a dog stick which can be let down to drag behind the wagon when travelling uphill. When not in use this is kept in place by a ring attached from a short length of chain to the rear cross-bar.

## (2) RUTLAND WAGON

Northwards from Hertfordshire through Bedfordshire and Cambridgeshire, the wagons become gradually larger and heavier. The wide double-railed sideboards give way to narrower single-railed types, the narrow wheels to wider straked tyres, while the small spindled frontboards are replaced by the deep, highly decorated frontboards of the wagons of Rutland and the adjacent counties.

Throughout this vast province, stretching from Bedfordshire to North Leicestershire and from Cambridgeshire to Warwickshire, the wagons are remarkably alike in design and construction. As there is some variation in detail, the most typical example, the Rutland wagon, will be described.

The Rutland wagon is a medium sized, double mid-railed wagon with a slightly curved profile and is used in the following counties:

BEDFORDSHIRE. The type of wagon gradually changes from the true Hertfordshire type in the south of the county to the Rutland type in the north of the county.

CAMBRIDGESHIRE. To the west of the Gog Magog hills the Rutland wagon is found. To the east of these hills, wagons of the East Anglian type are used, while in the south of the county the Hertfordshire type of vehicle is common.

HUNTINGDONSHIRE. The Rutland wagon is found throughout Huntingdonshire and occasionally in the Isle of Ely.

NORTHAMPTONSHIRE. Rutland wagons occur throughout the county, except in the southern corner where wagons of the spindle-sided bow type are used.

RUTLAND. Wagons are still very widely used on Rutland farms, and although towards the eastern border the large waisted Lincolnshire wagon is found, the smaller East Midlands wagon is also found throughout the county. On the larger East Rutland farms both types of wagon occur in great numbers. There all the wagons, irrespective of type, are painted brick-red.

LEICESTERSHIRE. The Rutland wagon is used throughout the county except in the Vale of Belvoir, where Lincolnshire wagons are found. In West Leicestershire, where the farms are smaller than in the east of the county, wagons are rare.

WARWICKSHIRE. The Rutland wagon is common in the south and east of the county. It penetrates westwards into the Vale of Evesham, where it is gradually replaced by the Worcestershire type.

While the Rutland wagon is considerably larger than the Hertford wagon, it is smaller than the Lincoln, and because of its almost horizontal top-rails, possesses none of the elegant appearance of the latter. The vehicle is well constructed and, owing to the small wheels and deep double mid-railed body, has a low, heavy appearance.

## WHEELS AND AXLES

The wheels are tyred with a single line of hoops of medium width, varying with the type of soil on which the vehicle is used. The normal tyre width is between 3 and 4 inches and 5 inches is rarely exceeded even on the heaviest clay in parts of South Warwickshire.

While wagons in the major part of the area were equipped with iron axles, or at least with iron axle arms, from as early as 1845, in Rutland all-wooden axles persisted until the present century and no wagon with iron axles was seen anywhere in the county.

In the same way, straked wheels persisted in Rutland until recent times, although the other counties had long before changed to hoop tyres.

## UNDERCARRIAGE

The undercarriage consists of two straight outside hounds running from a slightly curved iron-lined slider bar to the wide splinter bar at the front. On some wagons two inside hounds run the whole length of the forecarriage, on others they run only from the splinter bar to a point just behind the fore-axle. In all cases a pair of shutters are morticed at right angles to the hounds, giving the whole forecarriage added strength.

The pole joining the forecarriage to the rear axle is straight and ends in a chamfered edge just behind the axle, while both the pillows and bolster are wide and shallow. In order to decrease the height of the floor the bolster and pillows are no more than 4 inches deep and 7 inches wide.

## FRAMEWORK

This varies tremendously from one district to the next, and while the older wagons of pre-1850 date are equipped with simple straight side-frames, those of post-1850 date have waisted frames. This waisting varies from the slight waist of wagons in Rutland, Leicester, Warwick and Northampton to the fully waisted frame of wagons in Huntingdon, Cambridge and Bedford. Undoubtedly this technique was copied from the East Anglian wagons, although it did not appear in these southern East Midlands counties until 1880. In the north, wagons of post-1850 date and in the south all wagons built between 1850 and 1880 have the slightly waisted framework.[4] In this case the two forward side frames curve inwards and the front ends of the rear side frames are firmly bolted to it. Each frame is $3\frac{1}{2}$ inches square, while the main cross-bar is a straight block of beech 4 inches wide and 3 inches deep. Two inside summers run from the forebridge to the rear cross-bar and in some cases oak keys for fitting long boards join the summers and frames.

## BODY

The body is built with a series of narrow planks separated by two mid-rails. The lower side plank is constructed in two sections; the front section curves with the line of the forward side frame as far as the waist, while the rear section is perfectly straight from the waist to the rear cross-bar.

The two top-rails curve very gently and the body is the same depth throughout the whole length of the wagon. A series of wooden or, later, iron spindles join the side frame built of 3 inch square timber to the inner top-rail, usually of 2 inch square timber. The sideboards merely consist of the outer top-rail which runs parallel to the curve of the body. The space between the two top-rails is never filled. In some areas, especially in North Leicestershire, there are wagons which are of East Midlands type in general design and construction, but possess the removable harvest frames of the Lincolnshire type. These vehicles represent the transition between the true Lincolnshire wagon and the true East Midland wagon. In the Vale of Evesham the sideboards of the wagons continue beyond the line of the wagon to form fore-ladders, much in the same manner as on a Shropshire vehicle. On these wagons a pair of iron supporting brackets join the mid-rail of the frontboard to the front of the ladder.

The side supports, five on either side of the wagon, are mostly of wood. The main side support runs from a bolt on the inner top-rail to the two mid-rails and thence to the main cross-bar where it is morticed at its lower end. The other side supports are basically of the same shape and design, but are simply bolted to

30   Rutland Wagon from Teigh

the side frame. A number of iron brackets join the upper mid-rail to the outer side-rail and act as sideboard supports.

The frontboard is deep and highly decorated and, where it is joined by the mid-rails, has pieces of finely shaped sheet iron nailed to it. This ironwork, on the frontboard, is always formed in a fleur-de-lis pattern. The tailboard is removable and hinges on to the rear cross-bar by a pair of barrel eyes. While the wagons are often fitted with short fore-ladders, whose butt ends fit underneath one of the iron supports joining the two top-rails, they are rarely equipped with tail ladders.

COLOURS

In such a large area there is some variation in the colour of the body, and even more in the decoration of the frontboard. In Rutland, Huntingdon, Cambridge and South Northampton, the wagons are always painted a bright ochre red, while in Leicestershire, Warwickshire and northern Northamptonshire they are blue. In Bedfordshire the red colour gives place to dark brown and buff like the Hertfordshire wagon.

In Huntingdon, Cambridge and the major part of Northampton the plain, unornamented surface of the frontboard is inscribed with the name and address of the owner painted in black, but in Rutland, Leicester, North Northampton and Warwick, the frontboards are elaborately decorated. In addition to the black fleur-de-lis designs at the junction of mid-rails and frontboard an oval plate is painted at the front. Inscribed on its black surface is the name and address of the owner in white paint, while surrounding the plate is a complicated leaf design much in the Lincolnshire tradition. At the bottom the date of building and the name of the builder of the wagon is given.

In Leicestershire and Warwickshire frontboards are much plainer, sometimes with the name and address of the owner, sometimes without any inscription at all.

## BRAKING

The Rutland wagon is always equipped with a chain which can be passed around one of the rear wheels for travelling downhill. A few are also equipped with dog-sticks.

## 3 WEST MIDLANDS WAGONS

Wagons in the West Midlands of England vary tremendously from one county to the next, and there are far more sub-types in this region than elsewhere in the country. Even so, all types of wagon have numerous constructional features in common and one is quite justified in calling the West Midlands a single region. The main features of the wagon are:

Broad, greatly-dished wheels, usually shod with a double line of strakes.

A fairly deep body, painted blue or yellow, with one or two mid-rails.

Where sideboards are found they are narrow and boarded in. The side supports are elaborately shaped, usually from wood.

In profile the top-rails are gently curving, rising gradually from the centre to a deep frontboard and tailboard. The side frames are straight and the lock limited. The wagons of the West Midlands can be divided into three main types, (a) Southern type, found in Hereford and the adjacent counties, (b) Central type, found in Shropshire and Montgomery-shire, (c) Northern type, found in a wide belt stretching from the Stafford-Leicester borders to the North Wales Coastal Plain.

### (a) SOUTHERN TYPE

The heavy medium-sized wagons of Hereford and the adjacent counties may be divided into five sub-regional types, (1) Hereford panel-sided wagon, (2) Hereford plank-sided wagon, (3) Worcester-shire wagon, (4) Monmouthshire wagon, (5) Radnorshire wagon.

### (1) HEREFORD PANEL-SIDED WAGON

The traditional panel-sided box wagon is found throughout the county of Hereford from the Welsh border to the Malvern Hills. Unlike other regions, however, wagons are used for a variety of farm transport in addition to harvesting hay and corn. They are often used for carrying root crops and the produce of the cider orchards. The wagon is very solidly built, and during this survey vehicles built in the 1840s were seen in constant use in the county. It is said that the Hereford wagon is expected to last at least a hundred years, and this is not surprising considering its build, which is extremely heavy for what is, after all, a medium-sized wagon.

#### WHEELS AND AXLES

The wheels of a Hereford wagon are always double-tyred and each wheel is equipped with either a double line of strakes or an outer line of strakes and an inner line of hoops. As in other West Midlands counties, the practice of straking continued until relatively recent times, and hoop tyres never replaced strakes completely in Hereford. Since there is a great deal of clay land in the region, broad wheels are always found, even on the later plank-sided wagons.

Just as straking persisted until a late period, so, too, wooden axles were used as long as village craftsmen built wagons. For this

31   Hereford Panel-Sided Wagon

reason the naves are barrel-shaped and large enough to accommodate the thick wooden axle arms. The gap in the surface of the nave for extracting the linch pin is not covered by a straw-packed chock of wood, but is in the form of an uncovered tapering slit in the surface of the nave.

UNDERCARRIAGE

The forecarriage is well constructed and strongly built to bear the heavy weight of the body. Two pairs of hounds run the whole length of the forecarriage from the stout splinter bar at the front to the slightly curved iron-plated slider bar at the back.

Owing to its weight, two or more horses are required to draw the wagon. For this reason the forecarriage always ends in a splinter bar to which are attached two pairs of shafts. In the early nineteenth century one writer[5] noted that the wagons, which could carry about $3\frac{1}{2}$ tons, were so heavy that six horses were needed to draw a loaded vehicle. These horses were harnessed in pairs to the shafts.

The pole coupling the forecarriage to the rear carriage is straight and heavy, measuring some 6 inches wide and 4 inches deep and it ends just behind the rear axle.

The bolster and pillows are both made from heavy blocks of beech measuring as much as 10 inches wide and 8 inches deep.

The lower edge of the fore-pillow and the upper edge of the bolster are carved to a convex shape, providing considerable springiness in the fore part of the wagon.

## FRAMEWORK

The framework is uncomplicated in design, but like the under-carriage is extremely well constructed. It is straight, and the lock of the wagon is consequently limited, the turn of the fore-wheels being rarely more than 20 degrees. The framework consists of two curving side pieces, about 4 inches square, passing through the main cross-bar to join the forebridge to the rear cross-bar. A thick iron brace fitted to the side-frames and overlying the main cross-bar provides additional support at the centre.

Parallel to the side-frames are a pair of summers 3 inches square which run the whole length of the wagon. At right angles to the summers are a series of oak keys which provide a basis for the long-boarded floor.

## BODY

The deep body is constructed with three narrow side planks, separated by two mid-rails. The mid-rails and inner top-rails are slightly curved and run parallel to the side frames, which are themselves curved upwards at the front to provide a little extra lock. In appearance the body therefore has a gently curving profile, with the rails running almost horizontally as far as the line of the fore-axle, then rising gradually to the frontboard. The depth of the body is constant throughout its length and the wagon has none of the high aspect of the Lincolnshire wagon nor, indeed, the more subdued loftiness of its close relative, the Shropshire wagon.

The three side planks of the wagon are supported by a number of flat wooden standards morticed into the frame and the inner top-rail. In front of the main cross-bar the standards slope slightly forward, behind the cross-bar they have a slight backward slant. Each standard is nailed to the elm side planks. The two mid-rails project beyond the line of the frontboard, but in addition two other projections equal in length to the mid-rails jut forward at the front. These have been carved at the front of each side plank to project beyond the line of the wagon. They are purely decorative but occur on all true Hereford wagons, and the feature may be regarded as an essential part of local vehicle design.

K

The side supports are always of wood, each vehicle being equipped with four supports that run the whole depth of the wagon on either side. Both the main and rear supports are morticed to the cross-bar, and each one curves to the two mid-rails and then passes upwards to be bolted between the inner and outer top-rails. The two intermediate supports are bolted to the frame and top-rails, and each is the same shape as the main support. In addition to the four main supports a number of others, generally six on either side of the wagon, are bolted to the upper mid-rail and inner top-rail. A sideboard support on each runs from the mid-rail to the outer top-rail.

The outer top-rail is no more than 2 inches away from the inner top-rail, and the space between is only large enough for the top of the side supports. Unlike all other regional types the space between these top-rails does not form overhanging sideboards. However, the depth of the body can be increased 5 or 6 inches by fitting removable boards to the top-rails. Each board is equipped with four or five projections which can be fastened between the top-rails by a series of hooks and pins. These removable sideboards are only fitted for carrying root crops or cider apples. When the vehicle is required for corn or hay harvesting they may be replaced by harvest frames similar in type to those used in Lincolnshire, but smaller. As an alternative to sideboards and harvest frames some wagons have straight poles fitting into rings at each corner on the inside of the vehicle.

The deep frontboard with a definite arched top-rail is an important distinguishing feature of the Hereford wagon. The tailboard of the same shape is held by a pair of barrel-eyes passing through hooks on the rear cross-bar.

COLOURS

Hereford wagons are generally blue, but in the north of the county where the influence of the Shropshire wagon is felt, yellow-painted wagons are occasionally found.

The frontboard is plain, but it is customary to fit a finely-shaped white-painted board to the upper section. This bears the name and address of the owner, as well as the date of building the wagon, painted in black with a great deal of line decoration. In addition, a large white-coloured strip is nailed to the front of the wagon between the two mid-rails. This bears the name and address of the owner in black paint and follows the upper curve of the wagon body.

## BRAKING

The wagon is equipped with a locking chain to prevent too rapid descent of steep hills and, because of this, the junction of the

32   Hereford Plank-Sided Wagon

spokes and felloes of the rear off-side wheel is plated with iron to prevent wear.

## (2)   HEREFORD PLANK-SIDED WAGON

There was a radical change in wagon design during the last quarter of the nineteenth century when the older panel- or spindle-sided vehicle was replaced by plank-sided wagons. In many parts of the country, for example in East Anglia, the plank-sided wagons which were greatly influenced by the Scotch cart were a completely new departure in wagon design and were not in any way related to their   panel-sided predecessor. In Hereford, however, the plank-sided wagon was closely related to the traditional type, and the absence of upright standards and mid-rails in the body are the only features that distinguish later wagons from the traditional type.

The heavy undercarriage, the broad wheels and the absence of true sideboards still persists in the later vehicles, but instead of the two mid-rails of the older wagons, the plank-sided vehicles are characterised by two deep grooves that run the whole length of the body. The wagon is invariably painted blue, but the grooves are picked out in red or white. It is interesting that these two grooves do not follow the junction of the side planks but the line of the main frame and top-rail. The grooves are purely decorative and represent the mid-rails of the older vehicle on a plain-sided wagon. In sharp contrast to the plank-sided wagons in other parts of the county the side supports are invariably made of wood, and there are usually four on either side of the wagon. The main side support is similar in many respects to that of the older wagon, being a heavy piece of wood morticed to the cross-bar and inner top-rail. Once again the wagon is not fitted with sideboards although a removable harvest frame may be attached. Near each corner of the wagon there is an iron socket into which a pole is fitted giving additional support to a load of hay or corn. In some cases the wooden harvest frame is fastened permanently to the wagon with a series of iron brackets.

The arched top-rails of the tailboards and frontboards are again characteristic of the plank-sided wagons, as they are of the traditional types.

### (3)  WORCESTERSHIRE  WAGON

The Worcestershire wagon is plank-sided with the wheel and undercarriage construction characteristic of the Hereford vehicle. It occurs throughout the county of Worcester except in the eastern corner where East Midlands spindle-sided box wagons are used and the north-western corner where the Shropshire wagon with its distinct front 'sill' is found. The features of body design may be summarised as follows:

In profile the top-rails of the wagon are straight and the outer top-rail is greatly chamfered.

The body is basically plank-sided with three or four widths of 6 inch planking running the whole length of the wagon. It is not equipped with mid-rails, but a number of upright standards, rarely more than ten on either side of the wagon, are morticed to the frame and inner top-rails. These upright standards show slight traces of the panelled construction of the Hereford wagon.

The side supports are of iron, but there are only two on either
side of the wagon, a main support and a rear support of
plainly forged U-shaped irons.

The wagon has narrow, flat sideboards, the space between the
top-rails being filled with a thick elm plank running the whole
length of the wagon. A number of straight iron brackets run

33   Worcestershire Wagon from Powick

from the outer top-rail to be bolted to the centre of each
body standard.

The front and back top-rails display the characteristic bow-
shape of the Hereford wagon, but both the rail and frontboards
are plain and undecorated. The whole wagon is usually
painted yellow, while a long strip of white boarding bearing
the name and address of the owner is nailed on the sides of
the body, much in the Hereford manner.

The Worcestershire wagon may be described as a late, degenerated variation on the Hereford wagon since it has the typical
wheel and undercarriage construction of the latter although its
body is not as well constructed. Within Worcestershire very few
wagons existed prior to 1870 and the only four-wheeled vehicles
found in the county before that date were the flat-bodied trolleys.[6]
The design of the trolley was taken over by the large-scale manufacturers in the last decade of the nineteenth century and, as such,

trolleys became a common sight in many parts of the country. But in Worcestershire trolleys were village-made and date back at least to the last quarter of the eighteenth century. Indeed, at the village of Clifton-on-Teme there was a trolley reputed to be two hundred years old and another at Callow End dating from 1796.

34    West Midlands Trolley from Presteigne

The trolley consists of a flat platform placed on a wagon-like undercarriage with four wheels. The older village-made variety can be distinguished from the later mass-produced type by the high loading platform and the large diameter of the wheels. Often all four wheels of the later trolleys are of the same size, but the rear wheels of the older variety are much larger than the fore-wheels and are broad or narrow-tracked according to the local soil conditions.

The body of the trolley is carried on a series of wooden blocks on the forecarriage and axle. In Worcestershire the trolleys are always broad wheeled, the wheels being tyred in many cases with a double line of strakes.

## (4)  MONMOUTHSHIRE WAGON

Like the Worcestershire, the Monmouthshire wagon is a late nineteenth-century panel-sided box wagon, similar to the Hereford wagon in wheel and undercarriage construction. It is found in the rural parts of North Monmouthshire, parts of north-western Gloucestershire and occasionally in the Forest of Dean.

Within this region a considerable variation exists in the body design of the wagons. While some are closely related to the traditional Hereford wagon, others, although they are box wagons, show many characteristics of the West of England panel-sided bow wagons. Despite the variation in detailed design and construction these are the main characteristics of the Monmouthshire wagon:

The body framework is straight and similar in all details to the Hereford wagon, but in some cases vehicles are equipped with slightly waisted frames to provide greater lock. This design is particularly popular in north-western Gloucestershire where each wagon has a frame constructed in two sections, a slightly curved fore-section which extends as far as the main cross-bar, and a rear section which is straight and extends from the main to the rear cross-bar. These are bolted together to form a waist no more than 2 inches deep.

Each side consists of two or three elm planks, and despite the depth of the body there are no mid-rails.

A series of wooden standards, generally numbering from six to nine on either side of the wagon, run from the inner top-rail to the frame giving the wagon a panelled appearance. These are either morticed or bolted to the frame and morticed to the top-rail. Each standard is as much as 3 inches wide and 2 inches thick. Since there are no more than two side supports, the upright standards give added strength to the body of the wagon, which is mostly used for heavy, non-harvest work.

The side supports are either of wood or iron. Where iron supports are found they are U-shaped and similar to those used on the bow wagons of the lower Severn valley combining the function of side and sideboard supports. When they are of wood, they consist merely of slightly curved pieces of wood morticed to the main and rear cross-bars and bolted to the inner top-rail.

Some wagons have the removable sideboards and removable harvest ladders of the Hereford wagon, but others, especially in the east and south of the region, are equipped with wide, solid sideboards. In profile the top-rails of the wagon curve to a deep front and tailboard.

Both the front and tailboards are in most cases plain and undecorated, but in the Forest of Dean and north-western Gloucestershire a decorated, lettered board is often found.

35   Monmouthshire Wagon from Tirley, Gloucestershire

Most of the wagons are painted yellow but, again towards the
Hereford borders, blue wagons are quite common.

## (5) RADNORSHIRE WAGON

Despite the hilly nature of the Radnor and North Brecon countryside, wagons are still widely used in the region. Radnorshire,
although geographically in Wales, is more closely related to the
English Midlands both linguistically and culturally. For this reason
the wagon is relatively common in the area.

The Radnorshire wagon displays many of the characteristics of
the Hereford wagon, but in design and construction it is definitely
a degenerate type, this degeneration becoming more accentuated
as one moves westwards into the heart of Wales. While a wagon
seen at Newchurch displayed many of the characteristics of the
Hereford vehicle and was closely related to the parent type, another
seen at Llanbister in the heart of Radnor, had a less marked
relationship to the Hereford wagon.

While the Radnor wagon is unlike the Hereford type in that it
is equipped with narrow-tyred wheels, like the Hereford it has a
heavy, strongly constructed undercarriage and a straight frame.
The following are the features in which it differs from the Hereford
wagon:

> Since wagons in Radnorshire are not used for harvesting hay
> and corn but for carrying heavy material ranging from stones
> to root crops, and for carrying loads of apples from Hereford

cider orchards, the framework is very strongly constructed so as to bear great weight. Parallel to the side frames four summers run from the forebridge to the rear cross-bar. In addition, oak keys for long-boarding are morticed into each summer and side frame.

The side of the wagon is built up of four elm planks, a mid-rail running the whole length of the wagon separating the two pairs of planks. A number of upright standards are morticed into the frame, the mid-rail and inner top-rail, to give the vehicle a slightly panelled appearance. Two grooves, one above, the other below the mid-rail run the whole length of the wagon. As in the Hereford plank-sided wagon, these grooves do not follow the junctions of the side planks but the curve of the top-rails.

While the intermediate side supports are similar to those on the Hereford panel-sided wagon, the main side support is rather different. It consists of two pieces of wood bolted together to form a U-shape. The bottom of the U is morticed into the main cross-bar, its two extremities into the inner top-rail, the centre of each being bolted to the mid-rail. This firm side support gives added strength to the sides of a wagon designed to carry loose bulky material.

36　Radnorshire Wagon from Newchurch

The wagon has removable sideboards fitting into the narrow space between the top-rails. These are almost vertical and have the effect of deepening the body of the wagon by at least 6 inches. Since two-wheeled *gambos* and other carts are used for harvesting hay and corn, the Radnor wagon is not usually equipped with harvest ladders.

In design and decoration the frontboard is similar to that of the Hereford wagon, but since the Radnor wagon is not primarily a harvest wagon, the tailboard is rather different. Although most wagons are fitted with plain planked tailboards, they are also equipped with a special type for carrying root crops. This consists of a framework of curved iron spindles running from the top-rail to the bottom rail of the board and is fitted to the rear cross-bar through a pair of barrel eyes. Each side of the tailboard consists of two curved pieces of wood, bolted to each other at the top and bottom to form an oval shape. The whole tailboard can be let down at any angle to provide an extension of the wagon floor.

The Radnor wagon is painted prussian blue, and bears the name and address of the owner on a white board fitted at the front and also on a long strip of wood nailed to the side planks.

## (b) CENTRAL TYPE

The medium sized, panel-sided box wagons of Shropshire and the adjacent counties are lighter and more elegant than the wagons of Hereford. They may be divided into two sub-regional types, (1) Shropshire Wagon, (2) Montgomeryshire Wagon.

## (1) SHROPSHIRE WAGON

The Shropshire wagon is a panel-sided box wagon whose top-rails rise steeply from the centre to a lofty frontboard to give it a curved profile.

The wagon is found throughout the southern part of the county, but is absent from the north where a dairying economy predominates. South of a line drawn from Shrewsbury through Wellington to Newport, wagons are found on most of the farms, even though much of South Shropshire is hilly. The eastern boundary of the wagon is marked by the industrial belt of West Warwickshire. Southwards, to the Hereford boundary, the wagons become gradually heavier and more straight-lined, the under-carriages become more solid and all vehicles become far more closely related to the Hereford type. The Shropshire wagon extends up the Severn valley to the vicinity of Welshpool in Montgomery-shire, while it penetrates the Church Stoke gap as far as the village of Kerry. In the Welsh hills, to the west of Welshpool and Kerry, the wagons deteriorate in design and construction.

## WHEELS AND AXLES

The wheels of the Shropshire wagon are greatly dished, far more than those of any other type of box wagon. Since many Shropshire farms are located in valley bottoms and the arable extends up the slopes of the surrounding hills, the dish of wheels leads to greater stability when carrying loads. In addition it gives greater lock on a straight-framed wagon. Whereas the Hereford wagon requires as much as 52 feet for turning, the Shropshire wagon, despite a similar side construction, can be turned in approximately 36 feet.

As hoop tyres never replaced strakes in this region the wheels are always straked. Many of the farms are located in clay valleys and wheels are in most cases broad tracked with a double line of strakes. Each strake is fastened on with six or more broad-headed nails. One reason for the persistence of straking may be that the straked wheels grip the steep slopes far better than the hooped variety.

Although from an early date Shropshire had an iron industry within its boundaries, the products of the iron works were not used by local wheelwrights, for not only wooden supports on the body but also wooden axles persisted until a recent date. No wagon equipped with iron axle arms was seen anywhere in the county, and wagons built during the first quarter of the present century in the Coalbrookdale district itself were still equipped with heavy wooden axles and large barrel-shaped naves.

## UNDERCARRIAGE

The forecarriage consists of two outside hounds joined together by an iron-plated curved slider bar at the back. Two inside hounds run forward from the bolster and both inside and outside hounds are connected to the single pair of shafts by an iron shaft pin passing through the extremity of each hound. Shropshire wagons rarely have double shafts, but in the south of the county in the hilly neighbourhood of the Clun Forest, the characteristic heavy forecarriage and double shafts of the Hereford wagon make an appearance.

Both the bolster and pillows are wide but shallow, each being no more than 4 inches deep and 7 inches wide. The coupling pole is straight and ends just behind the rear axle.

## FRAMEWORK

The framework is very simple in construction and consists of two heavy side frames which curve gently upwards from the main cross-bar to the forebridge. To cut down weight the side frames, each about 3 inches square, are considerably chamfered. Running parallel to the side-frames are two inside summers 3 inches deep and 2 inches wide. Bolted to the underside of the forebridge, these pass over the main cross-bar and are bolted to the rear cross-bar. Since wagons in Shropshire are used for a great variety of farm purposes, long boarded floors are usual, and a number of strong oak keys between the longitudinal members act as a base for the floorboards.

## BODY

Both the top-rails and the single mid-rail of the body follow the curved line of the side frames. Each member is almost ski-shaped, being flat as far as the line of the main cross-bar and then sloping upwards at a sharp angle to the frontboard.

Joining the inner top-rail to the frame are a number of flat slats, each one morticed to the frame and top-rail and firmly nailed on the two side planks of the body.

The sideboards are solid and a single width of planking is sufficient to cover the space in between the top-rails. These sideboards project 8 or 9 inches beyond the line of the frontboards and provide an overhanging sill which acts as a seat for the driver and also as a support for the substantial fore-ladder. This ladder fits through a hook on the inner top-rail. When attached it has a gentle upward slope and is in effect a continuation of the curve of the wagon top-rails stretching forward some 2 feet beyond the front of the body.

The side supports are wooden and elaborately carved. Despite the early existence of an iron industry in Shropshire, wooden side supports were used long after other regions had changed to iron supports. In earlier examples of this type of wagon the main side support and the rear support are both Y-shaped, each made from two angled sections bolted together. The main and rear supports are morticed to the cross-bars and bolted to the mid-rail and the two top-rails. The four intermediate supports on either side of the body are also well shaped, and each one is bolted to the frame, the mid-rail and the outer top-rail.

The frontboard is plain and undecorated and, like the sides of the wagon, it is built with two planks separated by a single mid-rail. Three vertical slats are morticed to the forebridge and the front top-rail, while two supports are bolted to the forebridge, to the mid-rail and to the outer rail of the sill at the front. These are similar in shape but longer than the intermediate supports on the side of the body.

The tailboard is plain and like the frontboard it has a mid-rail and three upright ribs. It is attached to the rear cross-bar by a hinge and can be held open by means of a tie chain. Since wagons are used for a great variety of farm tasks in Shropshire, one of the main tasks being the harvesting of root crops, the plain tailboard can be replaced by a backboard with curved spindles running from the top-rail to the bottom frame of the board. The sides consist of two curved pieces of wood bolted together to form an oval shape. This construction is found in Radnorshire wagons and it also occurs in Montgomeryshire wagons, but does not occur elsewhere in the country.

COLOURS

The Shropshire wagon is always painted yellow and is without any decoration. It bears no inscription of any kind and there is never any indication of the owner or builder of the wagon. There is, however, the unusual custom of writing in pencil somewhere on the body the various duties performed by the wagon. The frontboard of a wagon seen at Bishops Castle was covered with pencil marks ranging from the blurred inscription 'Carrying swedes from Top Field 1914' to a very clear inscription 'Carrying hay from Lower Meadow 1952'. Year in, year out, this particular wagon was used for carrying hay, corn and root crops from the various fields of the holding and a clear picture could be obtained of the utilisation of the various fields over the course of thirty-eight years.

37   Shropshire Wagon from Plowden

BRAKING

The Shropshire wagon is rarely equipped with a roller scotch and never with a dog stick, but each has a braking chain which can be passed round one of the rear wheels for a downhill journey. For this reason one or both the rear wheels are plated with iron at a point where the spokes join the felloes, the iron plate encircling each spoke tip.

## (2)  MONTGOMERYSHIRE WAGON

The Montgomeryshire wagon may be regarded as a degenerate version of the Shropshire wagon. In the county it is a general-purpose farm vehicle used in the flatter parts of each holding mainly for stone and root crop carrying, and only occasionally for harvesting. Since the wagon is widely used for carrying stone and gravel, the side planks are much thicker than those of the Shropshire wagon, while the solidly built iron side supports give greater strength.

The simplification of design is related to the question of cheapness for the Welsh farmer in the past rarely had sufficient money for elaborate implements or vehicles. Peate[7], speaking of wagons in his native county, notes that the Montgomeryshire wagon must be looked upon 'as a deteriorated English Midlands type, the deterioration being caused on the one hand by its adaptation to heavy, non-harvest work, and by considerations of cheap building, and of the other by the fact that it occurs on the edge of the moorland, out of the traditional wagon-technique zone'.

The Montgomeryshire wagon differs from the Shropshire in the following points of construction:

Owing to the late development of the Montgomeryshire wagon, wagons pre-dating 1880 being very rare, the wheels have relatively slight dish, and many of them display the late nineteenth century construction of staggered spoke mortices. As in Shropshire, wagon wheels are always straked but they are often narrow.

The wagon is equipped both with a braking chain and dog stick, and some are also fitted with roller scotches.

The body consists of three or more thick side planks not separated by mid-rails, and the narrow sideboards are solid and continue beyond the frontboard of the wagon to form a sill.

The body itself is shallower than that of the Shropshire wagon

38   Montgomeryshire Wagon from Llanwnnog

and a few upright standards, generally six on either side, join the frame to the inner top-rail to give the wagon some vestige of a panelled appearance. These standards are morticed into the inner top-rail but only bolted to the frame, and are more in the nature of side supports than the ribs of a panel-sided wagon.

The two true side supports, one at the main cross-bar, the other at the back, are plainly forged U-shaped irons acting as supports for the sides as well as for the sideboards. In addition, straight iron brackets join the centre of each upright standard of the body to the outer top-rail.

## (c)  NORTHERN TYPE

In that they are broad-wheeled, panel-sided box wagons, the wagons of a wide belt of country stretching from the Leicestershire-Staffordshire borders to the North Wales coastal plain bear a close resemblance to those of Shropshire, except that the top-rails are almost horizontal and not curved. The wagons may be divided into two sub-regional types, (1) Staffordshire Wagon, (2) Denbighshire Wagon.

## (1)  STAFFORDSHIRE WAGON

The Staffordshire wagon is a heavy, panel- or sometimes spindle-sided box wagon with horizontal top-rails and broad, well dished wheels. Since much of Staffordshire is highly industrialised, only a few wagons still exist within the county, and all those seen during the course of this survey dated from the first half of the nineteenth century and were, with one exception, in a derelict condition. Because of lack of evidence it is rather difficult to determine the boundaries of the Staffordshire wagon but, in general, it may be said that the wagon occurs in the region extending from the West Leicestershire border, through the greater part of non-industrial Staffordshire and North Warwickshire to the Shropshire border and even to the Cheshire Plain. Although village-made wagons are not now found in Cheshire, the evidence does suggest that a few were known in the nineteenth century[8] and, since the degenerate wagons of Denbigh and Flint are similar in many respects to those of Staffordshire, it is probable that the Staffordshire wagon occurred in Cheshire in the nineteenth century.[9]

On the whole, the Staffordshire wagon is well constructed and well designed but, like all other West Midlands wagons, it is extremely heavy. In the words of William Pitt,[10] '. . . they are perhaps not capable of being improved . . . if anything may be condemned, they are rather too heavy, and should be constructed lighter if it could be done consistent with the giving to them a sufficient degree of strength'. In the early nineteenth century the broad-wheeled wagon was so heavy that it required six horses, while the narrow-wheeled vehicle required a draught power of four or five horses.[11]

## WHEELS AND AXLES

The Staffordshire wagon is equipped with narrow or broad wheels according to the type of country where it is expected to operate. Even the narrow single-tyred wheels are 4 inches wide, while the double-tyred broad wheels may be as much as 8 inches in width. In construction the wheels are similar to those of Shropshire wagons and are always straked and greatly dished. A wagon at Orton-on-the-Hill on the Leicestershire borders had wheels 5 inches out of the perpendicular, more dish than was seen on any other wheel during this survey. In addition, all the wagons seen in Staffordshire and the adjacent counties were equipped with thick wooden axles.

## UNDERCARRIAGE

The forecarriage is unusual in shape and consists of two straight

39 Staffordshire Wagon from Lichfield

outside hounds, each becoming smaller as it slopes back steeply from the axle to the straight, iron-plated slider bar. In addition, two inside hounds are bolted to the front of the outside hounds and curve backwards to a point just behind the fore-axle. These are joined by a straight shutter morticed into each, while the outside hounds are joined at the front by an iron shaft pin. Staffordshire wagons are seldom equipped with double shafts and so the forecarriage rarely ends in a splinter bar.

Both the bolster and pillows are made from stout blocks of wood 8 inches wide and 3 inches deep. These members are shallow and, consequently, the bed of the wagon is lower. The coupling pole, which measures 4 inches square, is straight and ends just behind the rear axle.

## FRAMEWORK

The framework is simple in construction and consists of a straight frame 4 inches square running from the forebridge to the rear cross-bar. In addition, two straight summers 4 inches by 3 inches run the whole length of the wagon. The main cross-bar is heavy and each side frame and summer passes over and is bolted to it. Since the frame is not waisted or notched, the wagon suffers from a limited lock. This is largely offset by the great dish of the wheels.

As the wagon is exclusively a harvest vehicle, it has cross-boarded floors and there are no keys joining the summers to the frame.

## BODY

The wagon is characterised by near-horizontal top-rails and has none of the lofty appearance of the Shropshire type. The depth of the body is almost constant throughout the length of the wagon, both frame and top-rails possessing an almost imperceptible curve down towards the centre.

The body itself is constructed with two widths of planking separated by a single mid-rail. A number of flat wooden standards, generally fourteen on either side of the wagon, are morticed into the frame, the mid-rail and inner top-rail. These standards are finely shaped, each being nailed to the side planks. One of the wagons seen had iron spindles instead of flat standards. The sideboards are much wider than is usual on West Midlands wagons and consist of a large number of wooden or iron spindles morticed to the two top-rails. Resting on these spindles is a single width of elm planking, nailed on with strips of sheet metal running from the inner top-rail, over the plank to the outer top-rail.

The main supports are wooden and vary from a type similar to those fitted on Shropshire wagons in the west of the region to those used on East Midlands wagons in the east. The more common type of side supports are shown in Fig. 24.

The intermediate side supports, invariably of wood, are bolted to the frame, while sideboard supports of iron run from the mid-rail to the outer top-rail.

In some cases, especially in the west of the region, the solid sideboards continue beyond the line of the frontboard to form a narrow sill above the front top-rail, as on Shropshire wagons.

The frontboard itself is plain and the name and address of the owner are only occasionally inscribed. In Fig. 25 a typical frontboard is shown. This is blue while a pair of iron struts running from the top-rail to the forebridge is black. The centre of the board is red in contrast to the blue of the body. Sometimes the name and address of the owner is written in blue or black. The backboard is equally plain. The two iron points of its supporting iron braces fit into a pair of staples on the rear bar. The backboard is therefore removable but cannot be adjusted in the usual way to the required distance from the body. It must be either tightly closed with a chained pin passing through the ends of the inner top-rails, or must be entirely removed. The wagon is generally equipped with fore-ladders similar to those used on Shropshire wagons, but there is no evidence to suggest that tail ladders were ever used.

## COLOURS

The wagon body is blue in the major part of the region, but yellow is used on the Shropshire-Warwickshire boundary.

## BRAKING

The Staffordshire wagon is equipped with a roller scotch, rarely with braking chains and never with dog sticks.

## (2) DENBIGHSHIRE WAGON

The heavy, straight-lined Denbighshire wagon is found on many of the larger farms of Denbigh and Flint. Its use extends westwards and a few wagons are still found on the farms of the North Wales

40   Denbighshire Wagon from Rhuthun

Coastal Plain, the Lower Conway Valley and even occasionally in the island of Anglesey. While, in the east, the wagons are fairly well constructed and in general design resemble those of Staffordshire, North Warwickshire and probably those of Cheshire as well, towards the west there is a distinct deterioration in style and construction.

The controlling factor in the design of the Denbighshire wagon is that it is intended for heavy non-harvest work and so the side planks, undercarriage and floorboards are much heavier and thicker than is normal on farm wagons. Except for its heavier construction the Denbighshire wagon shows many points of similarity with the Staffordshire wagon and differs only in the following:

> The two side frames are very heavy, each measuring 5 inches square, while the two inside summers are each 3 inches square.
> The various longitudinal members are joined by a number of stout oak keys on which are laid the long boards of the wagon floor.
> The degeneration of style and construction has resulted in the complete disappearance of a main cross-bar on some wagons but this weakness is offset by the existence of an extremely heavy framework.
> The fore-pillow and bolster are each bulged at the back to allow for the large king pin, 3 inches in diameter, which passes through them.

The body is built with three narrow side planks separated by two mid-rails, while a series of slats, morticed to the frame and top-rail are bolted or nailed to the planks and mid-rails. There are only ten of these ribs, five in front of the main cross-bar with a distinct forward slope, and five behind it with a distinct backward slope.

The wagon is expected to carry heavy material in bulk and the side supports are numerous and complicated in shape. There are seven on either side of the wagon, made either of wood or iron or of both materials.

The sideboards, approximately 10 inches wide, consist of an outer top-rail running parallel to the body of the wagon.

The body and floor of the wagon, although straight, have a distinct slope from back to front.

The tailboard is plain, fitted by means of two iron pins into staples on the rear cross-bar, but the frontboard is more complicated in design. The top-rail proper is arch-shaped like that of the Hereford wagon, but a heavy carefully-shaped block of wood is morticed to the projecting front ends of the side top-rails outside the front top-rail. This is carved to the shape of an ox yoke and acts as an additional support for the frontboard. The name of the owner, but nothing else, appears on this.

# 4 SOUTH-EASTERN WAGONS

In the counties of Sussex, Kent and East Surrey, the type of wagon used is a medium sized box vehicle, which has been described as the best of all English box wagons.[12] Its main characteristics are:

The wheels are of fairly small diameter, so that the floor of the wagon is low enough for easy loading. In the clay districts of the Weald the wagons are very broad-wheeled, but on the Downs narrow wheels are usual.

The body is fairly deep and panel-sided, with one mid-rail. The vertical slats are nailed to the side planks, but are not as numerous as in other panel-sided vehicles.

In profile the wagon has curved top-rails gradually rising from the centre to a plain frontboard and to an equally plain tailboard.

The sideboards are wider than is usual on box wagons and,

while the majority of sideboards are boarded, in East Surrey railed sideboards are common. Whatever their design, they rise at a steep angle from the inner top-rail to the outer rail. The main cross-bar is a solid block of timber, and all the forward side frames which are morticed into it curve sharply to provide a deep waist.

The wagons used throughout the area are remarkably uniform in design and construction, though there are minor sub-regional characteristics. The wagons of West Sussex are similar to those of the eastern part, but differ in having a removable body and an undercarriage which can be used as a timber wagon in a well-wooded district. Again, the wagons of Kent differ from those of Sussex in being painted a stone colour, while those of Sussex are blue. The south-eastern wagons may be divided into two types, (1) Sussex Wagon, (2) Kent Wagon.

## (1) SUSSEX WAGON

As has been said previously, the wagons of the major part of West Sussex differ from those of the east in having removable bodies and undercarriages that can be used as timber wagons. With this small difference the wagons of the whole of the county are remarkably uniform in design and construction.

### WHEELS AND AXLES

Marshall described the wagons of Sussex[13] as 'tall and large with a wide grasp or span between the wheels, which are here (in the Weald) frequently made with fellies six inches broad. Narrow wheels nevertheless are often in use. I have measured the ruts of a broad wheeled wagon, full six feet from out to out, or about five and a half feet from middle to middle, which is perhaps as good a width for farm carriages in general as can be fixed upon for a standard'. In the early nineteenth century oxen were still employed for drawing wagons in Sussex. These were 'driven by the goad and by the Yorkshire language'.[14]

The width of the tyre of Sussex wagons varies according to the type of country on which the vehicle is designed to operate. On the clay lands of the Weald broad wheeled wagons are common, while on the Chalk Downs narrow-wheeled vehicles are always found. While narrow wheels are generally tyred with a one-piece hoop, hooped wheels being post-1860, the majority of the broad-wheeled wagons are tyred with a double or sometimes a treble line of strakes. The width of these wheels varies from 6 inches to as much as 9 inches, and the relative ease of fitting strakes to broad wheels explains why straking persisted in the Weald for as long as wagons were built there.

As a general rule, most Sussex wagons are equipped with rear wheels no more than 4 feet 10 inches in diameter, a feature which gives a relatively low floor level. In addition, by gently curving the top-rails of the body the sideboards are made to clear the rear wheels, and there is no necessity for the complicated arched sideboard of West Country wagons. The use of low rear wheels completely avoids the loading difficulties of box wagons as well as the complicated arch design of bow wagons. On the other hand, the fore-wheels are large and have a diameter of up to 4 feet. In a deeply waisted wagon, such as the Sussex, large fore-wheels are quite possible since they do not interfere with the considerable lock of the vehicle.

Sussex is one of the most wooded of all English counties, and the availability of oak and beech for axle-making explains why so many Sussex wagons of the nineteenth century are fitted with wooden axles. Of course quite a number of wagons dating from the 1870s are equipped with iron axle arms carried in a wooden axle bed, as in other parts of the country. The hubs of wagons equipped with iron axle arms are much smaller, usually $12\frac{1}{2}$ inches in diameter and $12\frac{1}{2}$ inches long. The staggering of spoke mortices was also practised in Sussex from as early as 1870.

## UNDERCARRIAGE

The forecarriage of a Sussex wagon consists of two oak outside hounds joining a slightly curved, ash slider bar to the iron shaft pin at the front. There are also two inside hounds running from the front to just behind the fore-axle. Sussex wagons with a splinter bar for attaching two pairs of shafts are very rarely found, and if more than one horse is required to draw the wagon they are harnessed in tandem and not in line abreast. The pole joining the forecarriage to the rear carriage is slightly curved to clear the slider of the forecarriage and ends just behind the rear axle.

Both the bolsters and the pillow are plain and unchamfered, each being made of oak, ash or beech, 4 inches deep and 6 inches wide.

## FRAMEWORK

The outstanding characteristic of the Sussex wagon is its deeply waisted frame and body. The sides of all true Sussex wagons curve in at the front half of the body, and the average wagon can turn in a space of approximately 32 feet. In using curved side timbers Sussex craftsmen sacrifice the ideal form of framing found on Surrey wagons, which is four main timbers stretching over the undercarriage. The Sussex wainwright cuts his two curved forward side-frames from a thick oak slab already possessing a natural curve. Immediately above the fore-axle this slab is planed down to a depth of $3\frac{3}{4}$ inches and a width of $4\frac{3}{4}$ inches. These are tapered to a depth of no more than $2\frac{1}{2}$ inches at the rear end. The main cross-bar, morticed to take both the rear-end forward side-frames, consists of a solid piece of oak, often gently curving, 7 inches deep and 4 inches wide. This, as well as the rear cross-bar, is plated with a bar of iron along its upper surface. The forward side frames are morticed into the forebridge and the joints are reinforced by iron plates.

Most Sussex wagons are used for all farm purposes, for transporting shingle, stone, sand and root crops as well as for carrying the hay, corn and hop harvest. For this reason, wagons are equipped with long-boarded floors to provide a smooth surface for shovelling and have oak keys morticed into each longitudinal member of the frame. If, however, the wagon is specifically ordered as a 'Hay and Harvest Wagon', it is equipped with railed sideboards rather than the usual solid type, and the floorboards are laid crosswise.

## BODY

The sides of the body are built from two widths of plank separated by a mid-rail. The forward side planks following the curve of the side-frames are of poplar which is more flexible. These are cut very carefully for they must have the right curve so that they will spring into position. The rear side planks are elm, quite straight, each plank being 8 inches wide. The upper rear plank has to be wider at the rear end, the top surface having a slight curve to conform with that of the inner side rail under which it rests.

A number of flat wooden slats or ribs, each $1\frac{5}{8}$ inches wide and $\frac{5}{8}$ inch thick, act as supports for the side. There are up to fifteen of these slats on each side of the wagon morticed into the frame and inner top-rail and passing through the mid-rail. The side planks, $\frac{5}{8}$ inch thick, are fastened against these ribs on the inside of them. A thin nail with a wide head is used to hold the planking fast to the ribs. Each nail is clenched on the inside face of the plank and hammered flat.

The mid-rail runs back from the frontboard where it is slotted over the front mid-rail to a triangular piece of elm or ash which forms the waist. The rear mid-rail runs from the waist to the back of the wagon. The inner top-rail runs the whole length of the wagon, parallel to the rear mid-rail. Parallel to this runs the outer top-rail, cut so that the ends are 6 inches higher than the middle, which corresponds with the dip of the framework at the centre cross-bar. The inner top-rails are pierced by the inner legs of the iron side supports, which are held there by $\frac{9}{16}$ inch nuts.

The space between the inner and outer top-rails, which is at a sharper angle than usual, is completely filled in with ash plank some $\frac{3}{4}$ inch thick.

There are usually five supports on either side of the wagon. The main support and also the rear support are N-shaped, cut from $\frac{3}{4}$ inch to $\frac{7}{8}$ inch square iron bar, and welded to the required shape. Whether it is found on the centre cross-bar or the rear cross-bar the N-shaped iron is placed upright and across the wagon, the bottom ends of both legs are 7 inches apart and threaded to the beam.

41   Sussex Wagon from Horsham

Although iron supports are found in the newer wagons, the three intermediate supports are generally of wood, one above the fore-axle, the others in front and behind the rear axle. The frontboard is plain and consists of four or more ribs nailed to a pair of pillows separated by a mid-rail. The name and address of the owner are never written on the frontboard, but may be found on a plain board on the upper side plank at the front. Sometimes, especially in East Sussex, the wagons are equipped with a small wooden box fitted to the lower plank of the frontboard. This box carries the wagoner's food, perhaps a bottle of cold tea, and a number of tools and trace links in case one of the harness chains breaks. There is, too, a pair of large staples near each end of the front top-rail and a corresponding pair of smaller staples on the fore-bridge. Since the wagon is rarely equipped with fore- or tail ladders, poles 8 feet long are fitted to the staples at each extremity of the vehicle for use when carrying hay or corn. The tailboard is plain in design, built with two planks separated by a mid-rail and supported by a series of ribs. The board is fitted to the rear cross-bar by a hinge of barrel eyes fitting into bolts on the bar. This is removable and can be replaced by a pair of poles passing through staples at the end of the sideboards and on the rear cross-bar.

The wagon is also equipped with a rope roller in the form of a small winch on the rear cross-bar. This has two holes to take the ropes that are passed over the load and two round-headed pins for winding. It is held in place by two specially shaped eye-bolts of $1\frac{3}{4}$ inches diameter attached to the rear cross-bar. In addition, the wagon is equipped with a number of hooks located on the top-rails. Ropes thrown over the load of hay or corn are tied to these hooks in case the whole load tips on some of the steep slopes of the county.

## COLOURS
The body of the wagon is painted prussian blue, while the wheels, the undercarriage, the floor, and the inside of the wagon are all painted venetian red. There is rarely any painted or chamfered decoration on the wagon.

## BRAKING
The Sussex wagon is always equipped both with a drag shoe known in the area as a 'Skid-pan' and with an elm and iron roller scotch known as a 'Squat'.

## (2) KENT WAGON

In appearance and general construction the wagons of Kent and East Surrey are similar to those of Sussex, with the exception of the following details:

While the wagons of Sussex are generally blue, those of Kent are often painted a stone or buff colour.

While many Sussex wagons are broad-wheeled, in Kent broad

42   Kent Wagon

wheels are relatively rare and tyre widths of 3 inches or 4 inches are by far the most common.

Whereas the wagons of Sussex are never equipped with fore- and tail ladders, those of Kent are invariably fitted with them. The outer top-rail continues for some 12 inches beyond the line of the frontboard, the ends being joined by a stout piece of wood at right angles to this. An iron bracket sometimes joins the ladder to the frontboard. The tail ladder is similar to that found on Surrey wagons. It is short and nearly flat, acting as an extension to the floor of the wagon.

While on Sussex wagons rope rollers are almost invariably

found on the rear crossbar, in Kent they are not so common. In all other details the wagons of Kent, Sussex and East Surrey are remarkably uniform in size, design and construction. This uniformity throughout the south-eastern province of England is a notable feature of the 'transport geography' of our island.

## 5  WAGONS OF CENTRAL SOUTHERN ENGLAND

Although the wagons of Dorset, South Hampshire and West Surrey vary tremendously in size and construction from one district to the next, they all have a number of common features. The most important of these features is the fact that all the vehicles are spindle-sided box wagons. Those of the western sector of the region are the smallest of all English box wagons, while those of the eastern sector are medium sized and similar in many respects to those of Kent and Sussex. The wagons are characterised by the following features:

The wheels are of small diameter. In Dorset and South-West Hampshire they are very small indeed. Iron axles made an early appearance in this region, so that the larger hubs are not characteristic. The wheels are invariably narrow.

The body is shallow but, despite this, the wagons have one mid-rail. A large number of wooden or iron spindles join the frame to the inner top-rail.

In profile the top-rail of the wagon curves very gently so that the body is constant in depth throughout the length of the wagon.

The sideboards are wide and are either boarded in or consist of rails running parallel to the body. The side supports are either of wood or iron and join the inner top-rail to the frame and mid-rail. The bottom end of a wooden support is kept in place on the frame by half-round iron supports rather than by bolts. The frames are straight-sided and this limits the lock of the vehicle.

There are two main types of wagon in this province, (1) Surrey wagon, a medium-sized vehicle found in West Surrey and throughout Hampshire except in the north and west of the county,

(2) Dorset box wagon, found throughout Dorset, the eastern fringes of Devon and south-western Hampshire.

## (1) SURREY WAGON

The processes involved in building a so-called Surrey wagon have been described in great detail by George Sturt in *The Wheelwright's Shop*.[15] The wagon is a medium-sized box wagon usually painted light brown or buff.

Once again, as with so many other 'county' names, the term Surrey wagon is a misnomer. Not only is this type of wagon not used in much of Surrey, but it also extends to the western parts of Sussex, is found throughout southern and eastern Hampshire, and even penetrates South Berkshire.

In the east of the region there is no distinct boundary between the Surrey wagon and the south-eastern wagon, which is similar in many respects. Farther eastwards into Surrey the vehicles gradually change in design, deriving many features from both types of vehicle. At Dorking a transitional vehicle of this type was seen. In general outline this wagon was similar to the Kent wagon from which its shallow waist and solid sideboards are derived. On the other hand, its spindled body and numerous S-shaped side supports showed a relationship with the Surrey wagon. Farther south in Sussex in the village of Loxwood another transitional vehicle was seen. This vehicle, which was used on the same farm as a true Surrey wagon, had the panel-sided construction and solid sideboards of the Sussex wagon and, like the Sussex wagon, was painted blue. On the other hand, its unwaisted frame and S-shaped side supports show its close relationship with the Surrey type. Very roughly, the Surrey wagon may be said to exist west of a line drawn from Havant to Dorking, with extensions of the Kent type of wagon along the Downs to Guildford. In the north, of course, the influence of London is very pronounced and, since northern Surrey is largely urban, wagons are almost non-existent. Wagons are found in the greater part of Hampshire with the exception of the south-western part of the county where wagons are rare. In the north-western chalk area a bow wagon similar to that of Wiltshire is used. The Surrey type is common throughout eastern Hampshire and extends northwards almost as far as Reading in Berkshire.

## WHEELS AND AXLES

The wheels of the Surrey wagon are narrow, a tyre width of $2\frac{1}{2}$ or 3 inches being the most common. It seems that straked wheels were replaced by hoops at an early date in the region for, of the eighteen wagons of this type seen in Surrey and Hampshire, not one possessed

43   Surrey Wagon

straked wheels. The wheels themselves display some dish but, because staggered spoke mortices made an early appearance in the region, the dish is not very noticeable. Even wheels of pre-1870 date with no staggering of spoke mortices are not so greatly dished as the wheels of the Sussex and Cotswold wagons. One interesting feature of wheel design is the multiplicity of spokes; a rear wheel is equipped with fourteen spokes and a fore-wheel with twelve spokes.

Axles are generally of iron, but a few wagons with wooden axle beds and iron axle arms were seen. Not a single vehicle with all-wooden axles was found anywhere in the region. Because there are no wooden axles, the nave of the wheel is generally small, no more than 12 inches in diameter and 13 inches long.

In order to increase the lock on a straight-framed vehicle the axles are a few inches longer than is usual. The full length of an axle is 7 feet, leaving an inside axle width of 4 feet 6 inches, while the width of the body at floor level is 4 feet. The slightly dished wheels are set well away from the body thus greatly increasing the lock.

## UNDERCARRIAGE

The forecarriage of the wagon consists of the usual pair of outside hounds joining a splinter bar to the iron-lined curved slider bar. Unlike all other regional types, the Surrey vehicle always has a splinter bar at the front of the forecarriage, even though the vehicle has only one pair of shafts. These shafts are fastened to the bar by a series of barrel eyes. The hounds are morticed into the splinter bar and the junctions of the two members plated with sheet iron, which is nailed to the splinter bar and hounds. Both the bolster and pillow are deeper than is normal on wagons, each member being 8 inches deep. The main function of this is to raise the body of the wagon so as to increase the lock slightly. As the side frames are straight, the lock is small, but by dishing the wheels and increasing the depth of the bolster and pillow the lock is increased. Towards the end of the nineteenth century some wagons were built with fore-wheels of small diameter which could turn completely underneath the wagon. This necessitated a deeper axle-tree and a still deeper bolster. Frequently a second bolster similar to that on some Hertfordshire wagons was fitted above the axle-tree, the lower one being smaller, and separated from the upper bolster by blocks.

## FRAMEWORK

The body framework of the Surrey wagon has already been described (pages 89-90). A characteristic of the frame, however, is the thick iron plate overlying the centre cross-bar. The ends of this bar are sometimes drawn down and turned as hooks, on which the drag shoe is hung. The absence of a waist permits the use of continuous side frames and so the weight of timber is considerably less than that on waisted wagons. When seen in plan the frame and summers are straight, but in side elevation they curve upwards at their front ends. A groove was sometimes cut a little ahead of the main cross-bar where the fore-wheels rubbed on turning. This grooving gives the wagon a slightly waisted appearance since the lower side planks are also divided here to provide a very slight waist.

BODY

The sides of the wagon are built with ash planks, between $\frac{1}{2}$ inch and $\frac{5}{8}$ inch thick. There are two planks on each side separated by a mid-rail. This mid-rail and the two top-rails are of ash 2 inches wide by 1 inch deep. It is possible that the timber used for the rails is far too light and Sturt refers to broken rails being a frequent trouble with Surrey wagons.[16]

The side frame and inner top-rail are joined by a large number of round wooden or iron spindles passing through the mid-rail. When wooden spindles are used they are attached to the side planks by a number of screws, but on more recent wagons iron spindles of smaller diameter are fitted to the sides by staples hammered from outside and clinched on the inside of the wagon.

The sideboards are unusual in that the space between the top-rails is not filled with spindles or a solid plank, but with one or sometimes two rails that run from the front to the back of the wagon. These rails are of the same size as the top and mid-rails. Each top-rail is slightly curving, falling gradually from the front to its lowest level near the centre of the wagon, and then rising to the back of the wagon. The frontboard and tailboard of the vehicles are almost equal in depth, so that the vehicle has none of the high appearance of some other English box wagons.

Prior to the 1870s Surrey wagons were fitted with both wooden and iron side supports.[17] The main support running from the main cross-bar to the top-rails was invariably of iron, even on the earlier wagons. The design of this was simple and uncomplicated (Fig. 24). The rear support from the rear cross-bar to the top-rails is similar in shape to the main support, but the front support consists merely of a Y-shaped piece of iron joining the forebridge to the two top-rails. The intermediate supports of the earlier wagons were made of wood, generally four in number on either side of the wagon. Each one joins the centre of each to the outer top-rail. The side supports are kept in place at the bottom by a half round iron socket fitted to the floor of the wagon When iron side supports are used, they are curved and almost S-shaped, running from the frame to the mid-rail and thence to the outer top-rail.

The timbering of the frontboard is fitted outside the four supporting ribs in order to obtain a flat surface for the owner's name and address and the date of construction. The tailboard has a curved top-rail with ends rounded and trimmed down to a diameter of 2 inches. An iron collar is fitted near each end of the rail and about 12 inches of chain attached to it. The chain is fastened to a hook on the inner top-rail, so that the board can be carried horizontally if required. The board itself is undecorated, made of five strong ribs morticed into the curved top-rail and the straight lower rail. It is hinged, and is held on the rear cross-bar by a pair of barrel eyes fitting to hooks on the cross-bar.

The fore-ladder is straight and is set in an upright position with its lower ends resting against the forward face of the frontboard. An L-shaped bracket is bolted to the ladder 18 inches from the bottom and its angle hooks over the top-rail of the frontboard. The hind ladder, which is 6 feet long and quite straight, is held horizontally at the back by fixing the forward end of its legs into sockets on the side-frames.

Generally, as in south-eastern wagons, a rope roller is found at the back. Some wagons, unlike those of South-East England, have a rope roller not on the rear cross-bar but fitted to the underside of the tail ladder.

COLOURS
The Surrey wagon is painted a brown or buff colour with the undercarriage, wheels, floorboards and inner sides red. The frontboard is inscribed with the name and address of the owner of the wagon as well as with the date of construction. These inscriptions are painted in yellow or buff on the brownish surface of the board. The tailboard is plain and completely undecorated.

BRAKING
The Surrey wagon is equipped both with an elm roller scotch and a drag shoe. In Surrey and the adjacent counties only carts are equipped with dog sticks.

## (2) DORSET WAGON

Although the Dorset wagon is the smallest and lightest of all English box wagons, it is characterised by a very wide track.

The vehicle is found to-day throughout Dorset, parts of south-western Hampshire, the southern fringes of Somerset and the eastern part of Devon. In the early nineteenth century it was used over a much wider area of central southern England but, during

M

the nineteenth century, it was gradually supplanted in many areas
by the West Country panel-sided bow wagon. Writing in 1815,
Stevenson noted that it was widely used in Berkshire.[18] In construction the high Somerset wagon[19] shows many similarities to
the simpler and older Dorset box wagon. Indeed, even within the
county of Dorset itself the box wagon was gradually supplanted

44   Dorset Wagon from Sherborne

by a bow wagon in the last half of the nineteenth century.[20]
Despite this, there are still large numbers of box wagons in the
county.

In early nineteenth-century Dorset, according to Stevenson,[21]
the wheelwright's trade was a very extensive one. All the craftsmen
were engaged in wagon building, for at that time 'there were no
one-horse carts in the county, nor anything called a cart'.[22] To
carry manure, the Devon custom of using pack horses and dung
pots was employed, and the ubiquitous farm wagon was used for
all other farm transport. The reason for the great popularity of
wagons among Dorset farmers may be that the low wide four-
wheeled vehicle is more manageable than a cart on the long steep
hills that characterise Dorset.topography.

WHEELS AND AXLES

The wheels of the Dorset wagon are small in diameter and the
bed of the wagon low, making it easy to load. All the vehicles are
equipped with narrow tyres and even on the heavy clay of the

Vale of Blackmore wagons never have wheels more than 4 inches wide.

Iron axles, or at least iron axle arms carried in a wooden bed, made an early appearance in the county, and even wagons dating from the early 1840's are equipped with iron arms. For this reason the wheel naves are small, rarely more than 12 inches in diameter and 12½ inches long. These naves are almost cylindrical in shape, and the typical lemon-shaped naves of the bow wagon are unusual in Dorset. By 1870 the technique of staggering spoke mortices was widely practised in the county, consequently late nineteenth-century wheels display very little dish. Straked wheels also disappeared at an early date and, in the 1840's, it was customary to shoe wheels with one-piece hoops.

## UNDERCARRIAGE

Like the rest of the vehicle, the forecarriage of the Dorset wagon is small and lightly constructed. It consists of two outside hounds which run from an iron-plated curved slider bar to the front. Two inside hounds run from a hound shutter near the front to just behind the axle, while another inside hound runs from an iron hound shutter to the front. In most cases the shafts are fitted to the hounds by the iron shaft bar passing through the end of each. Even though the Dorset wagon is light, some vehicles are fitted with a splinter bar and two pairs of shafts, two horses being required to draw a loaded wagon up some of the steeper hills. Both the pillow and bolster are small and delicately shaped, each being 5 inches deep and 6 inches wide. The coupling pole is very short and curves upwards over the slider bar to end in a chamfered edge just behind the rear axle. The rear pillow is very small, no more than 3 inches deep and 5 inches wide. The short wheelbase of the Dorset wagon gives it exceptional lock and, although the frame of the body is not waisted or notched, the turning circle of the wagon is no more than 32 feet, which compares very favourably with some deeply waisted vehicles.

## FRAMEWORK

The framework of the body is very light, the side frames being no more than 2 inches square. Four straight summers run from the forebridge to the rear cross-bar each measuring no more than 1½ inches deep and 2 inches wide. As the wagon is used for all farm transport and not just for harvesting, its floor is long-boarded and laid on a number of oak keys which join the longitudinal members of the frame together.

## BODY

The frame, as well as the mid-rail and top-rail of the wagon, is curved, with the top-rails rising gently to a shallow frontboard and rather more steeply over the rear wheels to an equally shallow tailboard. The body is shallow but extremely strong for, in addition to a thick ash mid-rail, a large number of wooden spindles which pass through the mid-rail are morticed to the main frame and the inner top-rail.

The shallow body is built with an elm lower side plank no more than 7 inches deep which caps the side frame. This is followed by an ash mid-rail 1½ inches deep which, in its turn, is capped by an upper side plank 6 inches deep. The curved ash inner top-rail, of the same size as the outer top-rail, is 2 inches square. The nearly horizontal space between the top-rails is filled with a solid elm plank to provide fairly wide sideboards.

The side supports are wooden and there are usually five on each side of the vehicle. An important distinguishing feature of the Dorset wagon is that the bottom ends of the three central side supports, that is the main support and the two intermediate side supports next to it, fit into iron sockets. These carefully forged sockets are nailed to the frame and painted black.

For harvesting corn or hay the Dorset wagon is equipped with high fore and tail ladders similar to those fitted to the panel-sided bow wagons of Wiltshire.

## COLOURS

Most Dorset wagons are painted yellow but other colours, especially prussian blue or blue-black with red lines, are also used. In all cases the inside of the body and the under-parts are venetian red. The name and address of the owner is written both on the upper side plank and on the elaborately decorated frontboard, while on the tailboard appears the name and address of the builder and the date of construction. In many cases even the name of the painter and the date when the wagon was last painted appear. Both end boards are decorated with leaf and formal designs, making the Dorset wagon the most highly decorated of all English farm vehicles.

BRAKING

The wagon is never equipped with a drag shoe but it was customary to make the hoop of one or both of the rear wheels 2 inches thicker in one part. On going downhill the rear wheel is locked by a chain pinned to the main frame. The thicker part of the tyre is made to slide along the ground and provide a brake.

Each vehicle also has a small roller-scotch of elm which is hung from a pair of hooks on the rear axle when not in use. When it is required for going uphill one chain is wrapped around the wheel and hooked to the linch-pin of the rear off-side wheel so that the roller trundles behind that wheel.

The Dorset wagon may be described, therefore, as a square but light vehicle possessing a short, unusually wide body. Its great width and low build make it very stable on the hillside fields of the county. It is interesting that large scale manufacturers in the late nineteenth and early twentieth centuries produced great numbers of wagons based on the Dorset type and that mass-produced wagons, still known as 'Dorset Wagons', may be found in all parts of the country.[23]

# 6 YORKSHIRE WAGON

In north-eastern England wagons are limited to the East Riding and parts of the West Riding of Yorkshire. In Holderness broad-wheeled wagons are found on many farms, while in the Wolds, the Vale of Pickering and the other Dales of East Yorkshire, as well as on the Yorkshire Moors, wagons are widely used. They are less common in the Vale of York, while in the Tees Valley they give place to the large two-wheeled cart of the Scottish and North of England type. In the West Riding of Yorkshire the only four-wheeled farm vehicles are the flat trolleys which are used alongside the large Dales and Scotch carts. In the southern part of the Vale of York and throughout the non-industrial southern part of the county a few wagons of the Lincolnshire type occur. Throughout eastern Yorkshire, however, from the Humber to the Tees and from the Ouse to the coast, the wagons are remarkably uniform in design and construction and the only variation that occurs is in the size of the vehicles used in different parts of the county.

The Yorkshire wagon is unique in that it is neither a true box wagon nor a true bow wagon but displays characteristics of both types. It differs from all other regional types, and the only vehicle in any way similar to it is the late nineteenth-century plank-sided barge wagon common to much of the English countryside. These barge wagons were mass-produced in the last quarter of the nineteenth century and, as has been said previously, their design owed a great deal to the two-wheeled Scotch cart which was imported into England in the nineteenth century.

In 1641 Henry Best of Elswell, near Driffield, did not mention wagons in his *Family and Account Books*.[24] He did, however, mention 'wains' which, as far as one can gather from his account, were two-wheeled carts fitted with removable harvest ladders and possibly related to the Cornish wain, the Welsh *gambo* and the Scotch harvest cart.

Marshall, in 1788,[25] described the wain as 'a large ox-cart with an open body and furnished with shelvings, formerly used in carrying hay'. By the end of the eighteenth century it was being rapidly ousted by the new four-wheeled wagons for 'fifty years ago wains were pretty common, now there is not one perhaps left'.[26] In common with much of Highland Britain the north-east of England was the home of the Mediterranean ox-drawn cart, but in the late eighteenth and early nineteenth centuries these were gradually replaced by wagons for harvesting duties. Yet the replacement was not complete, for carts with harvest ladders and extremely long bodies are still widely used on the farms of East Yorkshire, and wagons are by no means as ubiquitous as Marshall leads us to believe. The Yorkshire wagon shows many characteristics of the older wains, especially in that up to the present time it is still equipped with a central draught pole and is obviously designed to be drawn by a pair of oxen. Its close relationship to the Scotch cart, on the other hand, is clearly shown by the plank sides and narrow horizontal sideboards.

N. A. Hudleston carried out a survey of Yorkshire wagons and published his results in 1952.[27] He notes the existence of three types of Yorkshire plank-sided wagons:

> THE DALES WAGON used in Bilsdale, Bransdale, Rosedale and other dales in North-East Yorkshire. This is a small, straight

framed wagon, only 8 feet long and 5 feet in maximum width. The rear wheels are 48 inches in diameter and the fore-wheels 26 inches in diameter. The wide solid sideboards, known as everings or shelvings, are almost flat.

THE MOORS WAGON used in the western part of the Vale of Pickering, in the coastal districts north of Scarborough and

45   Yorkshire Wagon from Carnaby

along the limestone hills. This wagon is slightly larger than that of the Dales, approximately 10 feet long and 5 feet in maximum width. The rear wheels are 56 inches in diameter and the fore-wheels about 38 inches. The wagon is equipped with solid sideboards which slope gradually towards the body and are not flat as on the Dales wagon. The Moor Wagon again is straight framed.

THE WOLDS WAGON used throughout the East Riding of Yorkshire and in the Vale of York. In Holderness the vehicle is equipped with broad wheels, while on the Wolds narrow wheels predominate. It is approximately 12 feet long and 6 feet in maximum width. The rear wheels are 60 inches in diameter while the fore-wheels are no more than 36 inches. Although the main frame is straight, on the later examples the fore-wheels swing right under the body as far as the pole to provide a considerable lock. The wagon is equipped with solid sideboards which slope more sharply than on the other two types.

Despite this three-fold division of the Yorkshire wagon, all the vehicles are similar in general shape and method of construction and sub-division is only justified on the basis of size.

## WHEELS AND AXLES

Except for some of the wagons in Holderness, all Yorkshire farm vehicles have narrow-tyred wheels. These are generally hooped, for hoops replaced strakes at an early date in the county.

Wooden axles, on the other hand, persisted until the 1880s and the iron axle was popularised only towards the end of the century when a Beverley firm of wagon builders began to supply the Yorkshire market. The persistence of wooden axles led to the continued use of large cylindrical wheel naves and the practice of staggering spoke mortices did not become popular until the early twentieth century.

## UNDERCARRIAGE

The forecarriage of the Yorkshire wagon is interesting and unusual in that two inside hounds, no more than 6 inches apart, project beyond the splinter-bar to provide a fixture for the draught pole. These two inside hounds, which project 8 inches beyond the splinter bar, diverge until they end just behind the fore-axle. The forecarriage which consists of the usual two outside hounds and curved iron-lined slider bar ends in a stout splinter bar. In addition to the projecting inside hounds, the splinter bar is fitted with a pair of iron brackets for attaching whipple trees, which are needed when a pair of horses is harnessed to the vehicle. Nineteenth-century writers favoured this form of harnessing, indeed some of them point out that the Yorkshire wagon has little to recommend itself except 'the mode of yoking . . . which is deserving of imitation'.[28]

A few of the later wagons are equipped with shafts attached directly to the splinter bar by sets of barrel eyes, but most of the vehicles have a central pole and whipple trees. Although the Yorkshire wagon is one of the lightest of all English farm vehicles, it is still the custom to use at least two horses abreast to draw it.

The pole joining the forecarriage to the rear carriage is slightly curved bending upwards over the slider bar and continuing backwards to just behind the rear axle. Both the pillows and bolsters are shallow, measuring no more than 4 inches deep, although on the full-lock wagons an attempt has been made to raise the floor of the vehicle by making the front bolster and pillow much deeper than is normal. In some cases the front bolster does not lie on the forecarriage as is usual, but is separated from it by a number of wooden pillars.

To prevent the fore-wheels from rubbing against the straight side frames on turning, the wagon is equipped with a pair of locking chains. Each chain runs from the rear end of the fore-axle to the side frames, a feature of construction noted by a number of early nineteenth-century writers.

## FRAMEWORK

The body framework of the Yorkshire wagon is very simple. The main frame and two summers run in a straight line from the front cross-bar, over the main cross-bar and are morticed into the rear cross-bar. Each member measures no more than 4 inches square but the side pieces display a considerable amount of chamfering. Indeed, various nineteenth-century writers criticised Yorkshire wagons for having too much decoration on the body. Strickland condemns Yorkshire craftsmen by saying,[29] 'in the making of both carts and wagons, too much attention is paid to fashion and the appearance of lightness, to the neglect of strength and utility. He is considered as the best wheelwright who can pare away the wood the smallest and who takes the greatest pains in carving the bodies of his vehicles to be painted bright blue and scarlet picked out with black'. On some of the quarter-lock vehicles the main frame is notched just where the fore-wheels would rub to provide a little extra lock. Since the wagons are primarily harvest vehicles, carts being used for all other farm purposes, cross-boarded floors are usual.

## BODY

The body is built with a series of planks without any supporting spindles or standards. In general there are three side planks 5 inches deep morticed into one another. The junction of each side plank is picked out in a different colour to the remainder of the body. The sideboards are solid and relatively wide, up to 10 inches, while the top-rails are slightly curved. The body is painted in a variety of colours, the most common being brown, with various members and lines being picked out in white.

On many wagons there are only two side supports, the first being an extremely plain iron construction over the main cross-bar. The rear side support above the rear cross-bar is equally plain. A few of the wagons have a front support consisting merely of an iron bar from the frame to the outer top-rail, but this is rarely found on any but the small Dales wagons. It is interesting that the four side supports found on each Yorkshire wagon are similar in shape and design to those found on the later plank-sided wagons of southern Britain; this is yet another feature pointing to the northern origins of those late, standardised wagons.

The frontboard is characterised by an almost ox-yoke-shaped top-rail, generally decorated with white lines. The board is built with a series of horizontal planks and a mid-rail above which a number of standards are found. The mid-rail and top-rail are considerably chamfered, the chamfers being picked out in white. The tailboard is of the same design as the frontboard and is attached to the rear cross-bar by a pair of barrel eyes.

COLOURS

Most Yorkshire wagons are painted brown, with buff or red undercarriage and wheels, although some vehicles are painted bright blue with red undercarriage and wheels. The line decorations on the frontboard are white.

BRAKING

The Yorkshire wagon is equipped with a roller scotch and drag shoe but rarely with a dog stick. This wagon may be described as a simply constructed plank-sided vehicle showing none of the fine workmanship of other four-wheeled wagons. One nineteenth-century writer[30] found little to say in favour of the wagons, 'they are high, narrow and long and are inconvenient for the purposes for which they are intended, namely carrying a top load'.

NOTES

1. LOWE, R. *General View of the Agriculture . . . of Nottingham*–London 1813–
   p. 18
2. [The hermaphrodite, usually shortened to morphrodite, morphey or morph, is a vehicle that is peculiar to the Eastern Counties of England. It is

found throughout East Anglia, Lincolnshire, the greater part of Nottinghamshire, parts of Derbyshire, the Vale of Belvoir in Leicestershire and even occasionally in Holderness. In a region traditionally engaged in crop raising on a large scale every available vehicle is needed for harvesting. For about ten months of the year this vehicle functions as an ordinary two-wheeled cart of the Scotch or tumbril type; for the duration of the harvest it is converted into a four-wheeled wagon by the addition of a forecarriage.

The hermaphrodite, known since at least the end of the eighteenth century, was described by Marshall in *Rural Economy of Norfolk*, 'With a common dung cart and a pair of old wagon shafts and forewheels, a carriage is formed which, partaking of both a cart and a wagon, is called a "morphrodite". The points of the shafts rest on the bolster of the forewheels to which they are fastened. A copse or foreladder, similar to that fixed on wagons but longer, is also supported by the bolster projecting over the horse in front.' These vehicles are light; they have a great lock and will hold as great a load of hay or corn as a wagon. Hermaphrodites may be divided into two broad groups.

(a) Those built from Scotch carts. The forecarriage of the hermaphrodite is fitted to the shafts of the cart some two-thirds of the way along its length, while an extra platform about 5 feet 9 inches long projects forward beyond the frontboard of the cart. This is bolted to the sides of the cart and supported at the front by two struts running to the forecarriage. The swivelling forecarriage itself is bolted to the shafts of the cart and a curved pole runs from the centre of the fore-axle to the tractive pin just in front of the cart body.

The hermaphrodite is equipped with a tail ladder fitting through the rear side support and resting on the rear cross-bar; it is also fitted with a fore-ladder fixed to the front end of the platform.

(b) Those built from tumbrils. In this case, since the shafts are removable, a slightly different type of forecarriage is used. This consists of a pair of wheels and an axle, with a cross piece fitted above the wheels. A pair of struts run from each end of this cross piece to the front of the carrying platform, while a straight pole runs from the centre of these to be pinned on the forebridge of the cart. Underneath this is fitted another pole running to the cart axle.

In many cases the carrying platform is not bolted to the cart body, but simply rests on two iron brackets at the sides. The coupling pole is therefore the only means of fitting the forecarriage to the cart.]

3. [In East Anglia the changeover from village-made, distinctly local vehicles, to standardized types was the result of influences not from the London area as in the remainder of England, but from Scotland. According to Arthur Young, *General View of the Agriculture of Essex*, 1813, Vol. 1, p. 161, carts of Scottish origin were imported by sea to East Anglia in the early nineteenth century. These carts were, without doubt, the primary influence in the development of the plank-sided wagon.]

4. MARSHALL, W. *Rural Economy of the Midland Counties*–London 1790– Vol. 1, pp. 136-7, '. . . the wagon is noticeable on account of its awkwardness, clumsiness and unwieldiness and in the present state of the roads its unfitness for a farmer's use. . . . No wonder it should require an acre of ground to turn in and a horse or two extra to draw it.' [Waisting was an attempt to overcome this great disadvantage that Marshall noted.]

5. DUNCOMB, J. *General View of the Agriculture . . . of Hereford*–London 1805–p. 45.

6. [In the following table the dimensions of two Worcestershire trolleys are ven.

|  | Trolley from Clifton on Teme Built c. 1750 | | Trolley from Callow End Built c. 1796 | |
| --- | --- | --- | --- | --- |
|  | Rear | Fore | Rear | Fore |
| **WHEELS** | | | | |
| Diameter . . . . | 52″ | 40″ | 54″ | 45″ |
| Method of Tyring . . | Strakes | Strakes | Double Strakes | Double Strakes |
| Width of Tyre . . . | 6″ | 6″ | 6¼″ | 6¼″ |
| Number of Spokes . . | 12 | 12 | 12 | 10 |
| Track . . . . | 58½″ | 66″ | 61″ | 60″ |
| **BODY** | | | | |
| Maximum Length . . | 139″ | | 144″ | |
| Maximum Width . . | 48″ | | 50″ | |
| Height of Body above ground | 50″ | | 55″ | |

7. PEATE, I. C. 'Some Aspects of Agricultural Transport in Wales', *Archaeologia Cambrensis*–December 1935–p. 234.
8. HOLLAND, H. *General View of the Agriculture . . . of Cheshire*–London 1808–p. 116
9. [Despite a very thorough search of Cheshire and South Lancashire no traditional wagon of any kind was found; indeed the majority of farmers expressed the opinion that wagons were unknown in the area.

Holland (*ibid.*–p. 117) does say that a few wagons were used in the county, especially in the arable district north of Northwich. These wagons must have been very rare, however, and they have now completely disappeared.

There are a number of reasons why wagons are so rare in Cheshire.
(i) The heavy drift soils are unsuitable for wagons and even the two-wheeled carts found in the county are generally equipped with broad wheels.
(ii) The economy of the county is dairying rather than arable and the two-wheeled carts are sufficient for leading hay.
(iii) Dispersed farmsteads are characteristic of Cheshire. 'The open field system seems to have been imperfectly developed here and the large nucleated village is a rarity. It is a county of small fields, small farms and a multiplicity of small villages and hamlets.' (Boon, E. P.–*Cheshire, The Land of Britain*, Part 65–Edited by L. D. Stamp–London 1941–p. 185). In 1844, Loudon wrote, 'Farms are very small; a great many under ten acres; only one or two at 350 or 400 acres. Excluding those under ten acres, the average of the county may be 70 acres.' (*An Encyclopaedia of Agriculture*–London 1844–p. 1164)

In Cheshire, however, a few mass-produced trolleys with small wheels are found, and occasionally a mass-produced plank-sided barge wagon occurs, but even on the largest farms no more than one of these vehicles is found. In Lancashire, too, wagons are extremely rare and, unlike Cheshire, there is no evidence to suggest that they were found in the early nineteenth century. Dickson, N. W.–*General View of the Agriculture . . . of Lancashire*–London 1815–p. 177 *et seq.* notes that there were no wagons or large carts in the county and that all farm transport was undertaken by light one- or two-horse carts similar to the Scotch cart. These are still widely used not only in Lancashire but also in Cheshire, North Shropshire and even in North-east Wales.]

10. PITT, W. *General View of the Agriculture . . . of Stafford*–London 1813–p. 47
11. *Ibid.*–p. 47
12. WAITING, H. R. and J. B. PASSMORE. *The English Farm Wagon*–

p. 20–MS in the Science Museum, dated 1936. 'The building of wagons seems to have developed earlier and to have been brought most nearly to perfection in the county of Sussex. Wagons which have been in use for three generations are still to be found in that county and their present state will stand as a testimonial to early workmanship.'

13. MARSHALL, W.  *Rural Economy of the Southern Counties*–London 1798– Vol. 2, p. 136

14. *Ibid.*–p. 134

15. STURT, G.  *The Wheelwright's Shop*–Cambridge 1923

16. *Ibid.*–p. 175 *et seq.*

17. *Ibid.*–p. 168

18. STEVENSON, W.  *General View of the Agriculture . . . of Dorset*–London 1815–p. 165

19. See Chapter V, Section 3(a)

20. See Chapter V, Section 2(d)

21. STEVENSON, W.  *op. cit.*–p. 165

22. *Ibid.*–p. 165

23. [A firm of agricultural engineers in Salisbury, Wiltshire, in the 1890s produced two types of vehicle. The first was called by them a 'Dorset wagon' and it can be regarded as a much simplified edition of the village-built vehicle. They called the second a 'Gloucestershire wagon', and this was very much a degenerate version of the South Midlands bow wagon, possessing none of the fine workmanship and the care for detail displayed in the village-built vehicle.

During the course of this survey, I saw at least twelve of these mass-produced Dorset wagons. Most of them were painted blue-black, and they could easily be distinguished from the traditional type by means of the following details of construction:

They were equipped with very small forewheels that swept right underneath the body. They were plank-sided without a midrail and, when they did have spindles, these were few in number; the side-supports were of iron and simple in design; the front and tailboards were entirely without decoration.]

24. BEST, H.  *Rural Economy in Yorkshire, being the Family and Account Books of Henry Best*–Surtees Society No. 33–1857

25. MARSHALL, W.  *Rural Economy of Yorkshire*–London 1788–Vol. II, p. 254

26. *Ibid.*–Vol. II, p. 254

27. HUDLESTON, N. A.  'Farm Wagons of North-East Yorkshire', *Transactions of the Yorkshire Dialect Society*–1952–Vol. LII, No. X, pp. 37-43

28. STRICKLAND, H. E.  *General View of the Agriculture . . . of the East Riding of Yorkshire*–York 1812–p. 82 *et seq.*–'The mode of yoking the wagons in a great part of the East Riding appears to be a practice peculiar to that district and is deserving of imitation. The four horses are yoked two abreast in the same manner as they are put to a coach, two drawing by the splinter bar and two by the pole; those at the wheel drawing also by a swinging bar, which the wheel horses of every carriage ought to do, as they thereby obtain considerable ease in their draft and are less liable to be galled by the collar than those which draw by a fixed bar. The driver then being mounted on the near-side wheel horse, directs the two leaders by a rein fixed to the outside of each of their bridles, they being coupled together by a strap passing from the inside of each of their bridles to the collar of the other horse. In this manner when empty they trot along the road with ease and expedition, and when loaded the horses, being near their work and conveniently placed for drawing, labour with much greater ease and effect than when placed at length. Were the wagons indeed of a better construction, the team would be excellent.'

29. *Ibid.*–p. 82

30. WALKER, G.  *Costumes of Yorkshire*–London 1885–p. 47 *et seq.*

# BOW WAGONS

## VI

## 1  SOUTH MIDLANDS SPINDLE-SIDED WAGON

From the western edge of the Cotswolds to the Chilterns, and from Warwickshire to Berkshire, the same type of wagon occurs with remarkable uniformity. This is a bow wagon of great elegance whose main characteristics are:

The large, greatly dished wheels. Where the wagon is designed for use on light soils, as in the Cotswolds, the Oxford Heights and the Chilterns, the wheels are narrow. Where it is used in the clay vales, such as the Vale of Oxford or the Vale of Aylesbury, broad wheels are usual.

The body is shallow and consists of a large number of wooden or iron spindles joining the inner top-rail to the frame. There is no mid-rail and the wagon is invariably painted bright yellow.

In profile the wagon is very elegant. The inner and outer top-rails are crooked, bending archwise over the rear wheels and rising at the back and front. The ends of the outer top-rails project backwards beyond the line of the body.

The sideboards are wide and consist of a large number of wooden or iron spindles joining the two top-rails. The distinctive side supports are made of iron.

The forwards side-frames are bent inwards to the main cross-bar providing a deep waist for the turning fore-wheels.

As the design and construction of the South Midlands wagon is so uniform throughout the province, no sub-regional types may be recognised. It is used in the following counties:

GLOUCESTERSHIRE. The South Midlands wagon is still common on the large farms of the Cotswolds. The western scarpface marks the western limit of the type, for in the Severn Valley panel-sided bow wagons predominate.

OXFORDSHIRE. The South Midlands wagon is used throughout the county, although in the dairying clay vales it is relatively rare.

WILTSHIRE. In the Vale of the White Horse and in north-western Wiltshire both the spindle-sided South Midlands wagon and the panel-sided Wiltshire wagon are found. In 'Cotswold Wiltshire' the South Midlands variety predominates, while in the Vale of the White Horse the Wiltshire type is most common.

WARWICKSHIRE. A larger version of the South Midlands wagon occurs in the southern limestone fringe of the county.

NORTHAMPTONSHIRE. In the southern fringe of the county the South Midlands wagon is found as far north as the Towcester district.

BUCKINGHAMSHIRE. The South Midlands type is found in much of the county, although it gives place to the East Midlands variety on its eastern fringes.

BERKSHIRE. The South Midlands wagon is common in East Berkshire, especially in the grain-growing district in the curve of the Thames between Maidenhead and Wargrave. It gradually gives place to the panel-sided bow wagon in the west of the county and to the Surrey type in th e south and east.

The South Midlands wagon, or the Cotswold wagon as it has been called, was looked upon by late eighteenth and nineteenth century writers on agriculture as the best of all English wagons. Marshall often advocated its use in all regions. In his *Rural Economy of Gloucestershire*[1] he noted that it was 'beyond all argument the best farm wagon I have seen in the Kingdom. I know not a district which might not profit by its introduction. Its most striking peculiarity is that of having a crooked side rail, bending archwise over the hind wheel. This lowers the general bed of the wagon without lessening the diameter of the wheels. The body is wide in proportion to its shallowness, and the wheels run 6 inches wider than those of Yorkshire wagons, whose side rail is 6 inches higher. Its advantages, therefore, in carrying a top load are obvious. And for a body load it is much the stiffest, best wagon I have seen. The price £20 to £25 according to the size and strength of the tyre. The weight is 15 cwts to a ton . . . Where and by whom it was first introduced I have not learned . . . it is by way of pre-eminence well entitled to the name of the farmer's wagon, for I have not seen another which compared with this, is fit for a farmer's use.'

Arthur Young considered the wagon worthy of a detailed measured drawing.[2] He called the vehicle a 'Woodstock Wagon', but it is substantially the same as the 'Gloucestershire Wagon described by Marshall and similar in size and construction to the wagons still found in the South Midlands province of England.

The outstanding characteristic of the South Midlands wagon is its elegance and general air of lightness. Despite its size, a narrow-wheeled South Midlands wagon weighs no more than a ton. This remarkable lightness has been achieved by shaving away all super-

46   South Midlands Spindle-Sided Wagon from Hailey, Oxfordshire

fluous weight in parts that are not liable to strain. The body of the wagon is painted yellow, a colour that adds considerably to the appearance of lightness. The various chamfers, of which there are a great number on these light wagons, are picked out in a different colour to the rest of the vehicle.

### WHEELS AND AXLES

The wheels, especially the rear wheels, are very large and so make arched sideboards necessary. In North Oxfordshire some wagons dating from the early years of the present century have much lower rear wheels. On these wagons, which are by no means traditional, the top-rails clear the top of the wheels with a gentle curve, similar to that found in Sussex wagons.

The fore-wheels, too, are larger than usual, but since the main frame and the body are waisted, large wheels with their corresponding advantages are quite feasible. The back of both the spokes and felloes is shaved greatly, but a considerable thickness of wood is left in those parts where the strain is greatest, especially at the junction of the spokes and felloes.

In the major part of the province iron axles made an early appearance and wagons dating from at least the 1830's are equipped with iron axles. For this reason the hubs of South Midlands wagons are distinctive, tapering from a diameter of as much as 14 inches at the hub mortices to as little as 3 inches at the linch-pin. The tapering is achieved by turning a series of concentric steppings in the surface of the hub, the steppings being painted black or yellow in sharp contrast to the red of the rest of the wheel.

This wagon usually has narrow wheels, except where it is designed for use on clay land. But the economy of the South Midlands clay vales based on pastoral farming with relatively small farms has limited the use of wagons in these areas.

Whether the wheels are shod with strakes or with hoops they display considerable dish, while the track is wider than on the majority of English wagons. Some authorities believe that the wide track of Gloucestershire wagons in particular was the prime reason why Brunel adopted a broad gauge track for the Great Western Railway.[3]

## UNDERCARRIAGE

Like the remainder of the vehicle, the forecarriage is very lightly constructed. It is simple in design, consisting of two tapering outside hounds joined together by a single shutter near the front. The thin, iron-plated slider bar is slightly curved.

Although wagons with a splinter bar for fitting two pairs of shafts are known in the clay vales and in Warwickshire, the most common method of shaft attachment is the round, iron shaft pin which passes through each end of the hounds. Double-shafted wagons are rare in the region.

Both the bolster and the pillow are shallow, no more than 4 inches deep and 6 inches wide. Each is greatly shaved with the draw knife and spokeshave to cut down excess weight. A locking chain joins the front bolster to the side frames to prevent the fore-wheels turning too far and rubbing against the thin side frame.

The pole joining the forecarriage to the rear carriage is straight and ends in a chamfered edge just behind the rear axle.

### FRAMEWORK

The framework of the body is complicated, for not only is the vehicle waisted, but the various members of the frame are rarely of constant dimensions. It is customary, even in the construction of the frame to shave away all superfluous weight. The main cross-bar consists of a piece of timber 3 inches square. This is notched on its upper surface to take the various framework timbers. The centre beam and the other members of the frame are completely below the floor line so that the floorboards are laid lengthwise parallel to the summers and side frames.

The floorboards are supported by cross-bars of heart of oak, $3\frac{1}{4}$ inches wide and $\frac{7}{8}$ inch thick. These are morticed through the summers about a foot apart. The floorboards, themselves of 1 inch thick planking, curve upwards to the front and conform with the curve of the frame. They are nailed to the cross-bars and rest so that their upper surfaces are flush with the upper surfaces of the summers and side-frames. Except for the slight projection of an iron bar above the cross-bar this arrangement provides a smooth surface for shovelling or sweeping. The method of constructing the main cross-bar saves considerable work in morticing, and consequently the timber can be much lighter than in many other wagons where the main cross-bar is greatly morticed.

### BODY

The body, which has no mid-rail, is supported by a large number of wooden or sometimes iron spindles, morticed into the frame and inner top-rail. A single elm plank is fitted to the rear side frame and each spindle is screwed into it. At the wheel arch a semi-circular plank and a curved top-rail are fitted to the main inner top-rail to make a bar. The front plank, which covers the forward side frame, curves inwards with the frame giving the wagon a deep waist. The waist and the great dish on the wheels enables the wagon to turn in a relatively small space.

The outer top-rail curves over the wheels, but, unlike that on the Wessex panel-sided wagons, the outer top-rail does not curve downwards to be made fast to the rear cross-bar. In fact it continues backwards and curves upwards. Ash is used for the top-rails because of its strength and flexibility. In addition it is often necessary to splice the outer top-rails just in front of the rear wheels, and ash is the best material for this.

It is customary to fill the wide space between the top-rails with a series of wooden or iron spindles but, in East Berkshire, railed sideboards are fitted with two rails running parallel to the top-rails, as on Surrey wagons. In the Vale of the White Horse and in Warwickshire the wagons have solid, planked sideboards, but throughout the remainder of the area spindled sideboards are customary.

The side supports are invariably of iron and there is no evidence to suggest that wooden supports were ever used. The form of these supports is distinctive, and there are four on either side of the vehicle. Like the rest of the body the frontboard is shallow, the top-rail being curved. A simple plank forms the board itself, which has the name and address of the owners and the date when the wagon was built written across it in black.

Although tailboards built with a series of spindles are found on some of the wagons, by far the most common type is a single, wide plank of ash or elm. This is reinforced at both ends by screwed or nailed cross pieces. The tailboard is rarely hinged but is fitted with two flat iron bars about $1\frac{1}{2}$ inches wide by $\frac{1}{2}$ inch thick, shaped to a round tongue at the lower end and fitting into stout square staples on the rear cross-bar.

The South Midlands wagon is never fitted with a tail ladder, but often has a short fore-ladder with hooked ends which clamp on the spindles of the sideboard.

### COLOURS

The South Midlands wagon is always painted yellow, while the wheels, the undercarriage and the inside of the body are red. On the frontboard, in black or red, is the name of the owner and on some wagons the address or village of origin as well. The tailboard is plain except perhaps for the initials of the owner.

### BRAKING

The South Midlands wagon is equipped with a roller scotch and drag shoe but never with a dog stick.

## 2   WESSEX AND THE LOWER SEVERN BASIN PANEL-SIDED WAGONS

Panel-sided bow wagons are used in a wide belt which stretches from Berkshire in the east to Glamorgan in the west. Although these vary considerably in detailed construction from one region to the next, a large number of features are common to all the sub-groups. The following are the main features:

The wheels are fairly small in diameter which lowers the floor and makes the vehicles easy to load.

The body, usually painted blue, is shallow but wide. The well carved vertical standards joining the inner top-rail to the main frame give the wagon an elegant panelled appearance. In profile the wagon has the top-rails bending archwise over the rear wheels. Very tall fore- and tail ladders are characteristic of the type and these are hooked to the floor of the wagon and are almost vertical.

The sideboards are wide, being widest at the arch over the wheels. They are usually boarded in, while the side supports are, more often than not, elaborately forged.

The side frames are straight and this limits the lock of the wagon.

The coupling pole of the undercarriage is curved and continues to the rear cross-bar where it is pinned.

When smaller wheels were introduced in the late-nineteenth century the design of all panel- and spindle-sided bow wagons changed considerably. Although the typical bow-shaped sideboards disappeared, many features of the older type were still retained. In the western limits of the bow wagon, notably Pembrokeshire and the Gower Peninsula, no true bow wagons are found, but the few wagons that do exist show many characteristics of the panel-sided variety (Fig. 53). These may be summarised as follows,

The body is panel-sided with numerous carved, flat standards joining the curved frame to the inner top-rail. Like many of their bow-sided predecessors the wagons are straight framed and not waisted. It is rarely that the wheels turn right under the body, and the wagons are generally equipped with iron wearing plates nailed to the frame.

The later wagons still preserve the wide sideboards of their predecessors. In profile, the top of the wagon is curved and the sideboard solid.

Although the side supports are not as beautifully designed as those of the bow wagons, they are invariably of iron and similar in general shape.

The high, almost vertical fore- and tail ladders of the panel-sided bow wagons were continued in these later types of wagon.

In South Wales the later wagons display the characteristic end boards of the older type.

Throughout the area the panel-sided wagon with a curved profile is found. This may be regarded as a degenerate and cheaper form of bow wagon which came into being as a result of the lessening of wheel diameters, and the consequent lack of need for the wheel bow.

There are six main sub-types of panel-sided bow wagons, (1) Wiltshire wagon, (2) West Berkshire wagon, (3) North-West Hampshire wagon, (4) Dorset bow wagon, (5) North Somerset and Vale of Berkeley wagon, (6) Glamorgan wagon.

## (1) WILTSHIRE WAGON

The Wiltshire wagon has the simplest construction of all the West of England bow wagons. It is light and elegant, although a number of writers on agriculture have condemned it as being too heavy for parts of Salisbury Plain with its abrupt changes of slope.[4]

The wagon is used throughout Wiltshire, although it is not so common in the northern half of the county where the economy is predominantly pastoral and the farms relatively small. In the south, however, where the larger farms have huge fields and the economy is based on arable farming, wagons are widely used and most farms possess at least four.

While the wagons are eminently suited to the topography of Salisbury Plain, they are less suited to the enclosed nature of the North Wiltshire landscape. The vehicle suffers from very poor lock and whereas this is not a disadvantage in the large fields of Salisbury Plain, it is a serious handicap in the northern section of the county.[5]

### WHEELS AND AXLES

The wheels of Wiltshire wagons are greatly dished, especially on wagons used in the chalk areas of the county. In those areas the

47  Wiltshire Wagon from Pewsham

wheels are narrow and a width of 2 or 3 inches is usual. On the other hand, in the clay vales of Wiltshire, wheels are broad with two lines of strakes on each wheel. It is interesting that iron axles made an early appearance in the county so that wagons dating from the second quarter of the nineteenth century often have iron axles. For this reason the hubs of the wagons are relatively small and cone-shaped, the diameter of each being 12 inches at the centre, tapering to 5 inches at the ends. The linch-pin is kept in place inside the hub by a chock of wood, which is fitted tightly by packing straw round its edges. A clasp and staple is fitted over the chock. The hub as a whole is delicately shaped and a number of distinct steppings turned by the lathe tools are found in concentric circles from the edge to the centre of the hub.

UNDERCARRIAGE

The forecarriage of the wagon is strongly constructed but an attempt has been made to cut away superfluous weight wherever possible. For this reason the hounds and slider bar of the fore-carriage are considerably shaved. The forecarriage consists of the usual pair of outside hounds running from the slightly curving slider bar to the front. There is a hound shutter just in front of the bolster. A pair of inside hounds morticed into this shutter run to the slider bar. The shafts are attached to the wagon by a shaft pin passing through the hounds. Double shafts, and consequently a wide splinter bar at the front of the forecarriage, are occasionally found, especially in the east of the county. In most of the area, however, double shafts are rare and when more than one horse is required to draw the vehicle they are harnessed in tandem rather than in pairs.

The pole joining the forecarriage to the rear carriage curves sharply over the slider bar and continues backwards and upwards from the rear axle to be fitted loosely on the rear cross frame by a keyed pin.

## FRAMEWORK

The body of the Wiltshire wagon is mounted high on the undercarriage. This technique, together with wide set wheels, helps to increase the small lock of the vehicle. In the same way the gentle upward curve of the side frames assists lock.

The body framework of the wagon is simple but strong and durable. The side-frames running from the forebridge to the rear cross-bar are straight, each measuring 4 inches square. The main cross-bar is much smaller than on waisted wagons and measures only $2\frac{1}{2}$ inches square. In addition to the three main cross-bars, there are two subsidiary cross-bars, one in front and the other just behind the rear pillow. These are nailed to the underside of the frame and, in addition to providing greater support for the frame, act as bases for a pair of side supports. Two inside summers, $3\frac{1}{2}$ inches deep and 2 inches wide, also run from the forebridge to the rear cross-bar. Since in Wiltshire the wagon is specifically a harvest vehicle long-boarded floors are rare, and wooden keys to join the summers and frame of the body are unnecessary.

## BODY

The sides of the body are panelled with a number of upright standards joining the side frames to the inner top-rail. The standards are not as numerous as on many panel-sided vehicles. On a wagon from the Vale of Pewsey there are twelve on either side. Each body standard, especially on the older wagons, is delicately carved to give the side of the wagon an elegant panelled appearance. The single side plank is held by two screws at the top and bottom of each standard. The inner top-rail curves upwards to the front top-rail and continues backwards to the tailboard. On the rear wheels the top-rail is, in effect, a mid-rail for a bow-shaped piece of timber is fitted to it. This is planked in with a semi-circular piece of wood to provide a wheel bow, the outstanding characteristic of all West Country wagons. The body itself is not deep and is boarded in with a single elm plank. The front part of the main side frames follows the curve of the inner top-rail, so that the wagon has a high appearance at the front. This feature also increases the lock of the vehicle.

The main side support joining the main cross-bar to the top-rails is a piece of delicately forged wrought iron. The base of this support is at the end of the cross-bar, but 3 inches of its lower end is concealed by a turned wooden spiral which acts as a washer between the cross-bar and the junction of the various members of each support. At this junction there is a hook to which the ropes, or wagon lines, are fitted, the hook being of a particularly fine design. The support itself is Y-shaped, one end running to the inner top-rail, the other to the outer top-rail. The rear support which runs from the rear cross-bar to the top-rail, is roughly the same shape as the main support, but since the sideboards at the back are considerably narrower, it does not have the same open appearance. The deep wooden washer is also missing. The front support, although it is fitted with a rigid and well designed iron hook, is simpler in design. It is morticed to the frame above the pillow and while one of its arms runs, parallel to the side plank, to the inner top-rail, the other falls away at an angle to the outer top-rail. In addition to the three iron supports on each side of the wagon there are two pairs of wooden supports. These run from either end of the subsidiary cross-bars to the inner top-rail, and thence to the inner rail of the wheel bar. Each is screwed to the frame and top-rails and acts as additional support to the bow.

A distinctive feature of all Wiltshire wagons is the wide side-boards. At the front these slope steeply in from the outer top-rail to the body; towards the back they become less sloping, and above the cross-bar they are almost flat. At the front they are relatively narrow, gradually increasing in width until at the wheel bow they are very wide indeed. Because of the greater weight the bow is expected to take there are two additional side supports here. While the inner top-rail continues in one piece from the front-board, the back board acting as a mid-rail at the wheel bow, the outer top-rail is in three sections. At the front the ends of the rail project upwards at a steep angle and the rail runs back as far as the line of the main cross-bar. Spliced to the rear end of this is the bow-shaped section which runs backwards in a curve to be morticed to the rear cross-bar. A slightly curved piece of wood is spliced to the bow itself continuing backwards to project beyond the line of the backboard. This exceptional construction of the outer top-rails on this type of wagon is a major feature distinguishing it from the South Midlands wagon where the bow does not continue downwards to the rear cross-bar.

The basis of the sideboards of the Wiltshire wagon is a number of iron and wooden spindles joining the two top-rails. The space between is filled with a single plank merely resting on the spindles. Strips of sheet iron are nailed to the top-rails on the planks of the sideboard to keep it in place.

The fore- and tail ladders are distinctive, for they are the tallest found on any English farm wagon. The fore-ladder is fitted at the front of the forebridge and continues upwards at a sharp angle 54 inches above the frontboard. It is kept in place by a pair of staples on the floor of the wagon, two hinged arms running from the ladder itself to be hooked on these staples. The rear ladder is shorter and rests on the backboard of the wagon. It does not rise at such a steep angle as the frontboard, but its ends are fitted in the same way, by a pair of staples on either side of the floor.

## COLOURS

The Wiltshire wagon is always painted blue. The frontboard, though undecorated, usually bears the name and address of the owner. The flat surface of the board is painted yellow with the inscription in black or blue. The tailboard is made of a single plank and has two hooks fitting into eye-bolts on the rear cross-bar. This is plain and undecorated and, like the rest of the body, is blue in colour. The under-parts and the inside of the body are painted red.

## BRAKING

The wagon is equipped with a roller scotch, a drag shoe and a chain to lock one of the rear wheels.

The general picture of the Wiltshire wagon is one of great elegance and sound construction. Its reduced lock makes it un-suitable for the small fields of the clay vales, but it is admirably adapted to the larger fields of the chalk downs and Salisbury Plain.

## (2) WEST BERKSHIRE WAGON

On the chalk downs of West Berkshire and East Wiltshire the most common vehicle is the narrow-wheeled, straight-framed wagon of Wiltshire type. However, there is also another vehicle which may

48   West Berkshire Wagon from Boxford

be regarded as an improvement on the Wiltshire type. With a waisted body-frame this type dates from between 1860 and 1880 and represents an attempt by the craftsmen of Berkshire and East Wiltshire to overcome the poor lock of the traditional Wiltshire wagon.

The chalk plateau of East Berkshire, which may be regarded as an extension of the Marlborough Downs, is a gently rolling stretch of downland, whose economy in the mid-nineteenth century was based on arable farming. The farms are large and so the large Berkshire wagon is common throughout the region. The northern boundary of the type is marked by the Vale of the White Horse, where the spindle-sided South Midlands wagon is found. To the south of the River Kennet the type is gradually replaced by the similar north-western Hampshire vehicle. The edge of the chalk plateau and the curve of the Thames between Wallingford and Goring marks the eastward extension of the wagon. It is interesting that in the eastern two-thirds of the county the wagons used are of either the South Midlands or the Surrey type. The table of measurements of a typical West Berkshire wagon shows that the vehicle is a larger version of the traditional Wiltshire wagon. It differs from the Wiltshire wagon in the following points of construction:

The West Berkshire wagon is heavier and larger and requires two horses to draw it. These are harnessed to two pairs of shafts fitted to a splinter bar on the forecarriage.

As on the Wiltshire wagon, the pole joining the forecarriage to the rear carriage curves over the metal-lined slider bar of the forecarriage. It continues backwards to be pinned on the rear cross-bar. An unusual feature is a small roller fitted in eye-bolts on the lower surface of the pole which prevents wear. Whenever the axles are greased the roller is also greased.

Like the Wiltshire wagon the West Berkshire vehicle is panel-sided with solid, arched sideboards bending over the rear wheels, but it differs from the former type in that it is a waisted vehicle. Long-boarded floors are also common on this type of wagon. The forward side frames are constructed from very strong, heavy pieces of timber, the usual size being 5 inches square. The fronts of these members are morticed to the forebridge and they curve backwards to be morticed into the main cross-bar. The rear side frames are usually thinner, about 3 inches square.

The wagon is equipped with a dog stick attached to the rear axle, while chains fitted to the forecarriage and side frames prevent the wagon from locking too far.

Apart from these four constructional differences the Berkshire wagon is similar to the Wiltshire type already described. It is interesting that this improved variety of the true Wiltshire wagon never became common in the area concerned. They were probably only made for some two decades in the late nineteenth century when cheaper mass-produced wagons were already making an appearance. The improved type did not completely oust the traditional quarter-lock vehicle, which is still quite common on the chalk downs of West Berkshire.

## (3) NORTH-WEST HAMPSHIRE WAGON

Just as wagons of the Wiltshire type extend into the chalk downs of Western Berkshire, so they extend into north-western Hampshire. In the chalklands of Hampshire panel-sided quarter-lock vehicles similar to those of Salisbury Plain, Dorset and South-West Hampshire are found in great numbers as well as the larger, half-lock vehicles. These vehicles, which are similar in many respects to the improved bow wagons used in West Berkshire, possess the characteristics of both the Wiltshire type of the wagons used in the major part of the Hampshire Basin and West Surrey.

As in Berkshire, however, this type of wagon is relatively rare, the main reason being that the products of two large ironworks, at Andover and Basingstoke, replaced the village-made wagons in northern Hampshire earlier than in any other area.

As may be seen from the table of dimensions, this wagon is almost the same size as that seen in West Berkshire, and all the wagons of both regions are identical in construction except in one point. The only difference between the two types is that while the Berkshire vehicle has solid, boarded sideboards of great width similar to the Wiltshire vehicle, those of north-western Hampshire have railed sideboards similar to the Surrey wagon. In this case the space between the main top-rail and the outer top-rail is filled by two inside rails which run parallel to the body of the wagon. The sideboards are wide, and over the rear wheels the wagon is as much as 85 inches wide. The detailed construction of the Hampshire wagon may be summarised as follows:

The sides are panelled and have a waist to increase the lock. The undercarriage and body framework are similar to the West Berkshire wagons. Long-boarding is usual.

49   North West Hampshire Wagon

The wagons have the very tall fore- and tail ladders of the Wiltshire wagon.

They are invariably narrow wheeled.

The iron side supports are similar to those on Wiltshire vehicles but, as in the West Berkshire vehicle, there are no subsidiary wooden supports near the wheel arch.

As on Berkshire wagons, the coupling pole is fitted with a greased roller, while the slider bar is lined with iron on its top surface. The pole itself is curving and is pinned to the rear cross-bar of the wagon.

The frame of the wagon is not usually fitted with iron plates against which the fore-wheels rub on locking. The forecarriage is, however, fitted with locking chains joining the frame to the carriage.

### (4) DORSET BOW WAGON

In addition to the small box wagons used throughout the county of Dorset, bow wagons similar to those of Salisbury Plain are also found throughout the county. Although these are less common than the box wagons, they were widely used, especially on the chalk lands of Cranborne Chase and Dorset Heights. The type also extends into South-West Hampshire as far eastwards as the River Test. It is interesting that in the Romsey district of Hampshire these bow wagons are known as 'Somerset wagons', which suggests their West Country origins. In south-western Hampshire and south-eastern Dorset wagons are rare since the negative nature of much of the New Forest and the smallness of the farms on its outskirts make wagons uneconomical and unnecessary. Throughout the region, however, both box wagons and bow wagons are found, the latter being far less common. As may be seen from the table of measurements, the dimensions of the Dorset bow wagon follow closely those of the South Wiltshire type. But while the shape and design of this wagon show close relationship to the Wiltshire type many details of construction suggest a similar kinship with the box wagons of Dorset. The Dorset bow wagon, therefore, represents the confluence of the Dorset box wagon and the true West Country panel-sided bow wagon. The characteristics of the Wiltshire bow wagon and the Dorset box wagon which are found together in the Dorset bow wagon are:

## WILTSHIRE BOW WAGON
### FEATURES

Shallow body with wide solid sideboards bending archwise over the rear wheels.

Panel-sided construction. Relatively few upright standards nailed to the side planks.

Y-shaped iron side and sideboard supports. Main support finely forged. Two wooden supports joining two subsidiary cross-bars to wheel bar.

High fore-ladder, held by a pair of staples on the floor of the wagon. Shorter tail ladder.

## DORSET BOX WAGON
### FEATURES

Single mid-rail; the body of two planks separated by mid-rail.

Forecarriage consisting of two outside hounds and unique construction of three inside hounds running from fore-axle to hound shutter.

Curved coupling pole ending just behind rear axle.

Body framework consisting of side frame and four summers.

Almost rectangular frontboard. Front top-rail very gently curving.

Boarded tailboard, with name of builder, his address and the date of the last painting inscribed. Both the frontboard and tailboard are elaborately painted and decorated with leaf designs.

## (5)  NORTH SOMERSET AND VALE OF BERKELEY WAGON

From Wessex the panel-sided bow-wagon may be traced along the Avon Valley into North Somerset and along the eastern shore of the Severn Estuary as far north as Gloucester. The extent of what may be termed the Vale of Berkeley wagon coincides with a natural region[6] consisting of the Lower Severn Valley and North Somerset as far south as the Mendips. South towards the Mendips the panel-sided wagon becomes rarer and is gradually replaced by the smaller south-western wagon. In Gloucestershire the eastern boundary of the wagon is marked by the steep scarp slope of the Cotswolds, while north of Gloucester there is an abrupt change from the Vale of Berkeley wagon to the Hereford box wagon in its various forms.

50  Dorset Bow Wagon from Plaitford

In the stretch of rich meadowland between the Cotswold scarp and the Severn, where most of the wagons occur, the basis of the economy is dairying. Though the farms are only medium-sized, each possesses its wagon, which is used mainly for harvesting the heavy hay crops as well as for the limited amount of corn grown.

The shape and design of the Vale of Berkeley wagon is similar to the Wiltshire type and differs only in the following points of construction:

Since much of the area is very heavy clay land, broad wheels are always found.[7] These are tyred with a double line of strakes or, sometimes, with an outer line of strakes and an inner hoop. Each strake is held by eight nails hammered at regular intervals into each strake. The wheels are also greatly dished, even more so than on a Wiltshire wagon.

Double shafts and the splinter bar construction are unknown. Although in North Somerset and in the southern section of the Vale of Berkeley iron side supports of the Wiltshire pattern are found on most of the wagons, northwards into the Vale iron side supports are replaced by stubby wooden supports, the Vale itself being particularly well wooded. The main side support which joins the main cross-bar to the inner side rail is held at the bottom by a half round iron bracket similar to those already described on the Dorset box wagon. In some cases a straight iron rod runs from the centre of this wooden support to the outer top-rail acting as a support for the sideboard. It is interesting that none of the complicated wrought iron main supports so characteristic of the Wiltshire wagon are found on those of West Gloucestershire. The subsidiary wooden supports, one just in front, the other just behind the rear wheels, are similar to those of Wiltshire. These are joined to subsidiary cross-bars under the body frame. The support above the front pillow and the rear side support are made of iron and are similar in design to those found on the Wiltshire wagon.

The end boards of the wagon are more highly decorated than

51    North Somerset and Vale of Berkeley Wagon from Dymock

those of the Wiltshire wagon. A semi-circular board bearing the name and address of the owner is fitted on the frontboard. This is highly decorated and its red or white background contrasts with the blue of the remainder of the body. The tailboard bears no inscription, but it is customary to carve or fit some form of design on its surface. On wagons in the Oldbury district a number of diagonal, crossed strips of wood are nailed on, and although the primary purpose of these strips is to give additional support to the tailboard, they also present a pleasing pattern.

### (6)  GLAMORGAN WAGON

From the Vale of Berkeley, the panel-sided bow wagon can be

traced along the western banks of the Severn, through South Monmouthshire to the Vale of Glamorgan. In a greatly degenerated form, it is found even in South Pembrokeshire, an area historically and culturally associated with Lowland Britain.

Although in shape and design the Glamorgan wagon is similar to the Wiltshire wagon and its varieties, in detailed construction it is somewhat different. Of all the panel-sided bow wagons the Glamorgan wagon is the most elegant and, according to Fox,[8] 'they possess the seemingly inevitable beauty and fitness of the last phases of the sailing ships and of other specialised creations which have been perfected by generations of men content to work in one tradition'. In Glamorgan they were regarded almost as family heirlooms and in many districts their first journey after delivery from the wheelwright was to take the family to church on Sunday. As in Gloucestershire, wagons were also widely used as hearses to carry the bodies of their deceased owners to the cemetery.

It is probable that the wagon type became fixed in the early years of the nineteenth century and the design was not changed until the early 1900s. Walter Davis[9] notes that bow wagons were used in many parts of South Wales in 1815 and he says that they were advantageous for carrying hay or corn from upland fields 'where carts with half the loads would hardly stand'.

WHEELS AND AXLES
Glamorgan wagons stand low and their wheels are small in diameter. Since they are not used in heavy clay land, the wheels are narrow and not greatly dished. Most of the wagons seen during this survey dated from the last quarter of the nineteenth century and all displayed the characteristic staggered spokes, small naves and iron-armed axles of that period.

UNDERCARRIAGE
The undercarriage of the Glamorgan wagon is lightly constructed. The forecarriage consists of four hounds, each outside hound being no more than 2 inches square, and each inside hound measuring 1 inch wide by 2 inches deep. These run from a slightly curved iron slider bar to an iron shaft pin passing through the extremity of each hound. There are rarely any hound shutters and never a splinter bar.

The coupling pole, 3 inches wide and 2 inches deep, is straight

o

52   Glamorgan Wagon from Llanishen

and, as on other panel-sided bow wagons, it continues beyond the line of the rear axle to be pinned to the rear cross-bar.

Both the bolster and pillows are light and shallow so that the floor of the wagon is not too high. Each one measures 6 inches wide and no more than $2\frac{1}{2}$ inches deep.

FRAMEWORK

The body framework is simply constructed and consists of two summers which run parallel to the side frames from the forebridge to the rear cross-bar. The vehicle is primarily used for harvesting and is equipped with a cross-boarded floor so that, apart from the cross-bars, there are no members at right angles between the summers. An attempt has been made to raise the floor of the wagon at the back by providing a subsidiary cross-bar over the rear pillow. As a result of this the wheel bow easily clears the rear wheels.

BODY

The body itself is built with a single side-plank having a mid-rail bolted on its outside edge to provide greater strength and support. The top-rails, which curve gently over the rear wheels, point upwards at the front, a feature which Fox has suggested[10] may represent the upward curve of the horns of an ox, while the balusters on the front and tail boards are the hairy fringe of the animal. Both features occur on all true Glamorgan wagons, even on those of the early twentieth century where wheel bows and high

53    Glamorgan Wagon from Pyle, a late example

rear wheels have given place to curved top-rails and lower wheels.

The sideboards are wide and, as in Wiltshire, they gradually widen from 15 inches at the front to as much as 19 inches at the bow. The two top-rails are joined by a series of wooden spindles morticed into each member. Resting on this framework is a single width of elm planking, kept in place by a number of sheet iron strips nailed above the planking.

The iron side supports are similar in shape to those found on Wiltshire wagons, although the forging seems to be more elaborate and the detailed decoration more carefully executed. The main side support with its greatly decorated base and elaborately curving rope hook is, in itself, a fine example of wrought iron work.

The frontboard, which is decorated, is capped by a number of round balusters morticed to the downward curving front top-rail and to a second arched top-rail above it. In the same way the plain tailboard also is capped by these balusters. The tailboard itself is attached to the wagon by two metal hooks fitting into a pair of staples on the rear cross-bar. The wagon is equipped with the high ladders typical of the Wiltshire and Vale of Berkeley variety.

COLOURS
The Glamorgan wagon is painted blue, with red under-parts. On the ribs of the frontboard is nailed a semi-circular painted board bearing the name and address of the owner and elaborate paintwork usually in black and white.

BRAKING

The wagon is equipped with both a roller scotch and drag shoe.

To summarise, the Glamorgan wagon is an extremely elegant and well constructed vehicle, of all bow wagons perhaps the nearest in design to the ox carts once so common in the western province of Britain.


# 3  SOUTH-WESTERN  WAGONS

The bow wagons of Somerset, Devon and Cornwall differ from the other bow wagons of the West of England in that the sideboards and top-rails do not continue backwards to complete the full half circle of the wheel arch, but end rather abruptly at the highest level of the arch. For this reason they are known as 'cock-raved wagons'. They do not have the elegant appearance of the wagons of Wessex the Lower Severn Basin and the South Midlands. In addition, most of the wagons are smaller in size. The main characteristics of the south-western wagons are:

The wheels are fairly small in diameter so that the floor level is low, making the vehicle easy to load.

The body is shallow at the front but, because of the wheel arch, it is very deep at the back. The body is either spindle-sided, this construction being particularly prevalent in Devon and Cornwall, or panel-sided, as with the majority of Somerset wagons. There is one mid-rail running the whole length of the wagon and another running from the line of the central cross-bar where the top-rails rise in the wheel arch to the back of the wagon.

The sideboards are wide and usually in the form of spindles, joining the two top-rails. Another thinner rail may run parallel to the top-rails from the front to the back of the wagon, giving additional support to the overhanging load.

In profile the wagon seems low at the front, but the steeply rising top-rails give it a lofty appearance at the back.

The side-frames are straight, and the lock is therefore limited. In Devon it is customary to make the fore-axle rather longer than is usual so that the fore-wheels stand well clear of the body. This greatly increases the lock. On a few Cornish wagons the fore-wheels are small and lock right underneath the body. Cornwall was the first area to adopt full lock vehicles as early as the first decade of the nineteenth century.

The south-western wagon represents the simplest form of bow wagon for, although it resembles the Dorset box wagon in that it is low and has one mid-rail, it also possesses many of the features of the more elaborate spindle- and panel-sided bow wagons. It has the high rear wheels, the arched sideboards and the tall fore-ladders of the Wessex and South Midlands wagons, and it may be regarded as a simplified form of those wagons.

In the huge region stretching from the Dorset borders to Lands End there are bound to be local variations in the style and design of farm wagons. Nevertheless there is, on the whole, a remarkable uniformity of construction throughout south-western England, the vehicles differing mainly in size. On this basis three sub-regional types of wagon may be recognised, (1) Somerset wagon, (2) Devon wagon, (3) Cornish wagon. It must be remembered that the boundary between these three types of vehicle is by no means clear-cut. The so-called Somerset wagon is found throughout south-eastern Devon and it only gradually gives place to the smaller Devon wagon towards the west. No distinct boundary can be drawn between the larger panel-sided Somerset wagon and the smaller spindle-sided Devon wagon.

## (1) SOMERSET WAGON

The Somerset wagon is the largest of the south-western bow wagons and is common throughout the county, except in the higher parts of Exmoor where two-wheeled carts are used for all farm purposes. To the north of the Mendips the south-western bow wagon gradually gives way to the Wessex variety though, even in the Avon Valley, both types of vehicle are found. The Somerset wagon also penetrates eastwards into the county of Dorset where it is seen occasionally alongside the older box wagon and the more recent Dorset bow wagon. It is found even as far east as the Isle of Purbeck, while westwards it is used along the South Devon coast until it is gradually replaced by the smaller Devon wagon.

### WHEELS AND AXLES

The wheels of a Somerset wagon vary in width according to the type of country on which the vehicle is expected to operate. In the Gordano district, for example, where the soil is heavy, wagon wheels are broad, up to 7 inches in width. On the lighter soils of the Mendips, on the other hand, 3 inch wheels predominate. Over the major part of the county, except for those districts on its western and eastern boundaries, the custom of tyring wheels with lines of strakes continued until recent times. Even during the present century, wheelwrights in the city of Bristol fitted strakes to wheels, although broad wheels were usually given an inner hoop and an outer line of strakes.

A certain amount of variation occurs in the design of wheel naves. In the north-east the naves are barrel-shaped with the characteristic turned steppings of the South Midlands wagon. These steppings are picked out in black, in contrast to the reddish-pink colour of the rest of the wheel. In the south-east of the county, on the other hand, wheel naves are cylindrical in shape and are similar to those fitted to Dorset wagons. In the remainder of the county, naves are lemon-shaped and have no stepped turnings on their surface.

Most of the wagons before 1870 have wooden axles but, around that date, the technique of making iron axle arms became widespread and, towards the close of the century, the all-iron axle was standard on most wagons.

### UNDERCARRIAGE

The forecarriage of the Somerset wagon is simple and light in construction. On the larger wagons it ends in a splinter bar to which two pairs of shafts may be fitted but, on the smaller variety common in western Somerset, the shafts are attached to the forecarriage by an iron shaft pin.

On the smaller wagons an attempt has been made to lessen the diameter of the fore-wheels and increase the lock by placing the fore-axle much lower without lowering the floor level of the vehicle. This is done by fitting a number of blocks between the bolster and the axle tree. Thus the fore-wheels are made smaller without any consequent decrease in the diameter of the rear wheels. While the ordinary double-shafted Somerset wagon needs as much as 40 feet to turn in, the lighter variety with low fore-axle can turn in a space of 28 feet.

The pillows and bolster are constructed from relatively light pieces of timber no more than 3 inches deep and 5 inches wide.

The coupling pole curves over the straight slider bar and con-

tinues beyond the line of the rear axle to be pinned to the rear
cross-bar.

## FRAMEWORK

The body framework, like the undercarriage, is lightly and simply

54 Somerset Wagon from Durston

constructed and similar in general design to that found on Dorset
box wagons. Since the side frames are straight there is less need to
use heavy pieces of timber in their construction, and thus the frame
consists of two straight pieces, slightly rounded in cross-section
with a diameter of 3 inches, which run from under the forebridge
to a mortice on the rear cross-bar. On some wagons the main
cross-bar is very light, consisting of a piece of timber 2 inches wide
and 1 inch deep joining the two main side supports. This is nailed

to the body summers. On many wagons the cross-bar is 2 inches in diameter and round in section rather than square. Again it is bolted underneath the summers. The main side supports are morticed into each end of this cross-bar. The rear cross-bar is square in cross-section, but equally light.

In Somerset two-wheeled carts are used for all farm transport except harvest work, and for this reason it is only rarely that wagons are equipped with long-boarded floors. Thus there is no necessity for the usual keys to take long boarding.

## BODY

Each side body is constructed from two planks separated by a single mid-rail, sometimes round in cross-section, which runs the whole length of the wagon. The top-rails of the wagon curve over the large rear wheels and continue almost in a straight line to the back of the wagon. Another ash rail is bolted to the top-rail just in front of the rear wheels and is bent upwards in a bow over the wheels. The space between this bowed top-rail and the lower inner rail is filled with a semi-circular plank.

The wagon is more often than not panel-sided, a number of flat standards running from the inner top-rail to the frame. These are nailed or bolted to the side planks and pass through the mid-rails. But, in addition to panel-sided wagons, there are many spindle-sided vehicles in Somerset. These are especially prevalent in the south of the county and also towards the Devon boundary. The close relationship of the spindle-sided Somerset wagon and the Dorset box wagon is apparent in the construction of the body. The Dorset wagon with its shallow body, single mid-rail and spindles, is obviously the south-western spindle-sided wagon without the wheel arch and the consequent lofty appearance of the rear. On the spindle-sided Somerset wagon, however, each spindle is wired to the side planks with a piece of wire passing through the planks from the inside of the wagon and twisted around each spindle.

Whether the wagon is spindle-sided or plank-sided, the sideboards are always spindled, often with a middle top-rail running parallel to the other top-rails.

The side supports are of wood and are never morticed either to the inner top-rail or to the side frame. Once again, as on Dorset wagons, there are two or three iron sockets on either side of the wagon holding the bottom of each intermediate side support. Each support is bolted both to the mid-rail and to the inner top-rail, while a bar of twisted metal runs from the mid-rail to the outer top-rail acting as a sideboard support. In some cases, especially on West Somerset wagons as on the Devon, the rear sideboard support runs the whole depth of the wagon from the outer top-rail to the rear cross-bar. This support, which is of spiralled iron, ends in a ring attached to the cross-bar.

The frontboard is often plain, although it may be inscribed with the name and address of the owner. Fig. 25 represents a typical frontboard, the shape of which varies little throughout Somerset and South-West England generally. In shape it is similar to those found on Dorset wagons, although the elaborate lettering and painted decorations of the Dorset type are rarely found on Somerset wagons. Unlike the Dorset wagons the tailboards are perfectly plain with no inscription. The two iron braces that run the whole depth of the board fit into a pair of staples on the rear cross-bar. Two short chains and iron pegs keep the board in place. The wagons are generally fitted with very high fore- and tail ladders similar to those on Wessex wagons. The bottom end of each ladder is hooked to a pair of staples on the floor of the wagon, while the ladder itself rests against the inside of each end board.

## COLOURS

The bodies of most Somerset wagons are painted blue although yellow wagons are not unknown. The undercarriage, wheels and interior of the body are painted red, with the iron-work in black, and the inscriptions and decorations either in white or yellow.

## BRAKING

The Somerset wagon is usually equipped with a chain to lock one of the rear wheels on a downhill journey. Sometimes they are equipped with drag shoes and roller scotches, but never with a dog stick.

## (2) DEVON WAGON

From Somerset the wagon occurs westwards along the coastal plain of South Devon, through the numerous valleys of Dartmoor to the north coast of the county. They are less common than in Somerset and are found only in the flatter parts of the county, becoming rare towards the Cornish boundary.

In the early nineteenth century, Devon wagons were rare and the few wagons in existence were found only on farms in the Axminster district and on a few of the farms of 'agricultural gentlemen'.[11] Marshall notes that wheeled vehicles of any kind were uncommon in Devon during his visit to the county,[12] although one or two wagons 'of west country construction' were seen in the southern part.[13]

In construction the Devon wagon, known in the county as a 'Ship wagon' owing to its high back, is similar to the Somerset type. It has the characteristic lemon-shaped naves, rising top-rails and straight framework, the main difference being in size rather than construction. Devon wagons are small and light, but most are equipped with rope rollers fitted either to the rear cross-bar or to one of the horizontal members of the tail ladder. On the steep hillside slopes of Devon a rope roller is essential to secure the load firmly.

Both in Somerset and Devon the introduction of smaller wheels with staggered spoke mortices brought about a radical change in the late nineteenth and early twentieth centuries. Much smaller rear wheels were then fitted, the wheel arch became unnecessary,

55   Devon Wagon from Upton

and top-rails were made straight or slightly curving. Despite this change, many traditional features of design still persisted so that later south-western wagons with a single mid-rail and numerous spindles have the appearance of a Dorset box wagon. Many of them have fully locking forewheels and most are painted blue-black and are thus easily distinguishable from the Dorset wagon.

Although the craftsmen of towns such as Bristol, Taunton and Okehampton adopted this new design, most of the craftsmen clung to the traditional wagons, and as late as 1935 they still produced straight-framed bow wagons.

### (3) CORNISH WAGON

A few small wagons similar in construction to those of Devon are found in eastern Cornwall but, throughout the main part of the county, only a few wagons are found. The traditional harvest vehicle is the two-wheeled wain (Fig. 5) and the few wagons that occur are really four-wheeled versions of this.

In early nineteenth century Cornwall there was a great variety of agricultural vehicles, ranging from simple ox-drawn sledges to full-lock four-wheeled wagons. Each was adapted to the varying requirements of the Cornish countryside, and as Worgan says,[14] 'a hilly country like Cornwall excites exertion and is productive of much ingenuity'. Thus the hay wagon described by Worgan was a light, elegant vehicle with arched sideboards, similar in general construction to that of Devon (Fig. 55), but much smaller. The sides were low, and because of the enclosed nature of Cornwall the wagons were given a great lock by making the fore-wheels so small that they swung right under the body. This[15] is the earliest reference to full-lock vehicles and the existence of this construction in Cornwall pre-dates any other full-lock vehicles in England by at least fifty years.

Besides wagons of the bow type, another simpler four-wheeled wagon was common in Cornwall in the nineteenth century. This vehicle has a body consisting simply of a flat platform with fully locking wheels and a pair of arched guards over the rear wheels (Fig. 56). These guards were merely part of the floor of the wagon and the vehicle, except for the wheel guards, had the appearance of the common trolley (Fig. 34). This vehicle was definitely derived from the wain and could carry up to 300 sheaves of corn compared with the 250 sheaves carried by a wain.[16]

As the Cornish countryside is hilly, a load of corn or hay must be securely tied down. Thus, most of the harvest vehicles are equipped with rope rollers for tightening ropes passed over the load. On many vehicles there are two rollers fitted to the cross-bar of the cart or wagon. Both wains and wagons are fitted with fore- and tail ladders hooked to staples in the floor of the vehicle.

56   Cornish fully-locking Wagon

Sometimes the rope roller is fitted to the lower part of the tail ladder rather than to the rear cross-bar.

Despite the existence of a few wagons in Cornwall, the county must be regarded as part of the cart-using zone of Britain. The few wagons found there are merely small four-wheeled versions of the older carts.

## NOTES

1. MARSHALL, W. *Rural Economy of Gloucestershire*–Gloucester 1789–Vol. II, p. 35
2. YOUNG, A. *General View of the Agriculture . . . of Oxford*–London 1809–Plates 23-4
3. LANE, R. H. 'Waggons and their Ancestors', *Antiquity*–1935–Vol. IX, p. 144 *et seq.*
4. DAVIS, T. *General View of the Agriculture . . . of Wiltshire*–London 1813–p. 38–'The wagons are rather handsome but in general are too heavy . . . the wagon weighs from 19 cwt. to 22 cwt.'
5. MARSHALL, W. *Rural Economy of the Southern Counties*–London 1798–

p. 326–'The West Country wagon is common on these hills. It differs from that of the Cotswolds in having no insections in the body to receive the forewheels in turning. In an open country there is less occasion for such mode of construction than in narrow enclosures; and the body is not only more roomy and commodious, but is stronger by continuing the side pieces throughout from end to end. And for the road where heavy loads and long journeys are required, whole bodies have their advantage. But for harvest work in an enclosed country, insections are highly useful.'

6. JERVIS, W. W. 'The Lower Severn Basin and the Plain of Somerset', *Great Britain. Essays in Regional Geography*–Cambridge 1928–p. 110 *et seq.*

7. [Louis Hall of Town Farm, Oldbury, in an interview in August 1956, said that a wagon with broad wheels can be taken across a field immediately after rain and the broad wheels will leave no marks on the grass.]

8. FOX, C. 'Sleds, Carts and Wagons', *Antiquity*–1931–Vol. 5, p. 154

9. DAVIS, W. *General View of the Agriculture . . . of South Wales*–London 1815–Vol. I, p. 208

10. FOX, C. *op. cit.*–p. 155–'The use of oxen as draught animals in Britain renders it not improbable that the curved ends of the sideboards and the rising curve of the upper rail of the headboard between them are zoomorphic—reminiscent of the frontal of the horned ox; the balusters then represent the fringe of hair about his broad brow.'

11. VANCOUVER, C. *General View of the Agriculture . . . of Devon*–London 1808–p. 125

12. MARSHALL, W. *Rural Economy of the West of England*–London 1805–Vol. I, p. 117

13. *Ibid.*–Vol. I, p. 123

14. WORGAN, G. B. *General View of the Agriculture . . . of Cornwall*–London 1815–p. 37 *et seq.*

15. *Ibid.*–p. 37

16. *Ibid.*–p. 38

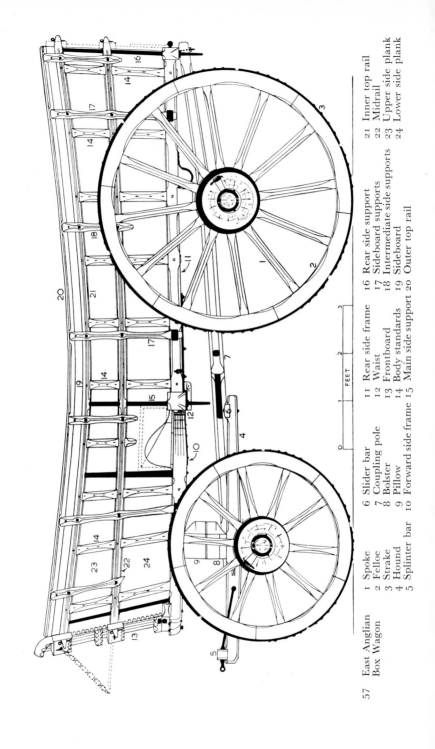

57 East Anglian Box Wagon

1  Spoke
2  Felloe
3  Strake
4  Hound
5  Splinter bar
6  Slider bar
7  Coupling pole
8  Bolster
9  Pillow
10 Forward side frame
11 Rear side frame
12 Waist
13 Frontboard
14 Body standards
15 Main side support
16 Rear side support
17 Sideboard supports
18 Intermediate side supports
19 Sideboard
20 Outer top rail
21 Inner top rail
22 Midrail
23 Upper side plank
24 Lower side plank

FEET

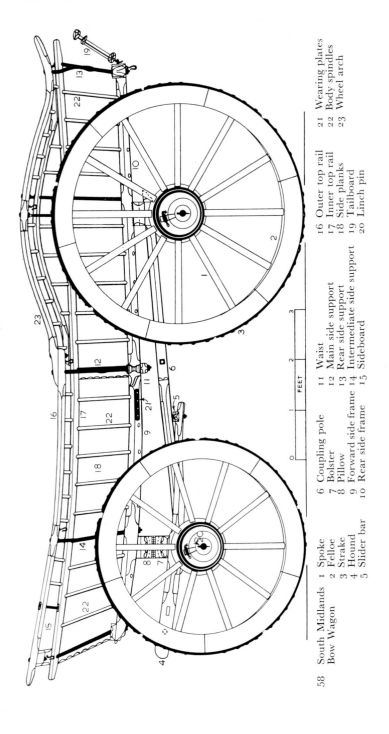

58 South Midlands Bow Wagon

| | |
|---|---|
| 1 Spoke | 6 Coupling pole | 11 Waist | 16 Outer top rail | 21 Wearing plates |
| 2 Felloe | 7 Bolster | 12 Main side support | 17 Inner top rail | 22 Body spindles |
| 3 Strake | 8 Pillow | 13 Rear side support | 18 Side planks | 23 Wheel arch |
| 4 Hound | 9 Forward side frame | 14 Intermediate side support | 19 Tailboard | |
| 5 Slider bar | 10 Rear side frame | 15 Sideboard | 20 Linch pin | |

FEET

0   1   2   3

# APPENDIX I

## MEASUREMENTS OF EXAMPLES
### OF EACH
## WAGON TYPE

The number of wagons measured were as follows:

### BOX WAGONS

1 EASTERN COUNTIES WAGONS
    (1) Lincolnshire Wagon . . . . . 22
    (2) East Anglian Wagon . . . . . 19

2 EAST MIDLANDS WAGONS
    (1) Hertfordshire Wagon . . . . . 15
    (2) Rutland Wagon . . . . . . 54

3 WEST MIDLANDS WAGONS
    (a) Southern Type . . . . . . 14
    (b) Central Type . . . . . . 18
    (c) Northern Type . . . . . . 12*

4 SOUTH-EASTERN WAGONS
    (1) Sussex Wagon . . . . . . 44
    (2) Kent Wagon . . . . . . 10

5 CENTRAL SOUTHERN ENGLAND WAGONS
    (1) Surrey Wagon . . . . . . 18
    (2) Dorset Wagon . . . . . . 36

6 YORKSHIRE WAGONS . . . . . . . 24

### BOW WAGONS

1 SOUTH MIDLANDS SPINDLE-SIDED WAGONS . . 58

2 WESSEX AND THE LOWER SEVERN BASIN PANEL-
    SIDED WAGONS
    (1) Wiltshire Wagon . . . . . . 24
    (2) West Berkshire Wagon . . . . . 8
    (3) North-West Hampshire Wagon . . . 4
    (4) Dorset Bow Wagon . . . . . 8
    (5) North Somerset and Vale of Berkeley Wagon . 14
    (6) Glamorgan Wagon . . . . . 9

3 SOUTH-WESTERN WAGONS
    (1) Somerset Wagon . . . . . . 24
    (2) Devon Wagon . . . . . . 6
    (3) Cornish Wagon . . . . . . 7

* Nine were in a derelict condition and all the measurements could not be ascertained.

P

## BOX WAGONS—1 EASTERN COUNTIES

### (1) LINCOLNSHIRE WAGON

| | Leadenham, Kesteven. Built 1829 | | Sleaford, Kesteven. Built 1915 | | Cotham, Nottinghamshire. Built c. 1860 | |
|---|---|---|---|---|---|---|
| **WHEELS** | REAR | FORE | REAR | FORE | REAR | FORE |
| Diameter . . . . | 61″ | 52″ | 60″ | 48″ | 61″ | 50″ |
| Method of Tyring . . . | Hoop | Hoop | Hoop | Hoop | Hoop | Hoop |
| Number of Strakes . . | — | — | — | — | — | — |
| Width of Tyre . . . | 3″ | 3″ | 3″ | 3″ | 3″ | 3″ |
| Number of Spokes . . . | 12 | 10 | 14 | 12 | 12 | 10 |
| Track of Wheels . . . | 63″ | 63″ | 64″ | 64″ | 62″ | 62″ |
| **FRONT** | | | | | | |
| Maximum Width (Top) . . | 62″ | | 48″ | | 54″ | |
| Depth of Frontboard . . | 33″ | | 33″ | | 34″ | |
| Width of Sideboard . . | — | | — | | — | |
| Width of Forebridge . . | 46″ | | 43″ | | 43″ | |
| **BODY** | | | | | | |
| Total Length (Top) . . | 146″ | | 144″ | | 146″ | |
| ,, ,, (Bottom) . . | 120″ | | 120″ | | 122″ | |
| Front (Top) to Ground . . | 72″ | | 78″ | | 74″ | |
| Mid Point (Top) to Ground . | 64″ | | 76″ | | 68″ | |
| Rear (Top) to Ground . . | 74″ | | 72″ | | 76″ | |
| Front (Bottom) to Ground . | 39″ | | 42″ | | 40″ | |
| Mid Point (Bottom) to Ground . | 41″ | | 42″ | | 35″ | |
| Rear (Bottom) to Ground . | 44″ | | 42″ | | 40″ | |
| **COLOURS** | Prussian blue Buff undercarriage White decoration | | Red Buff undercarriage Black decoration | | Prussian blue Red undercarriage White decoration | |

## BOX WAGONS—2 EAST MIDLANDS

### (1) HERTFORDSHIRE WAGON

| | Barley, nr. Royston. Built c. 1890 | | Bishops Stortford. Built c. 1860 | | Harpenden, Road wagon. Built in early 1900s | |
|---|---|---|---|---|---|---|
| **WHEELS** | REAR | FORE | REAR | FORE | REAR | FORE |
| Diameter . . . . | 62″ | 42″ | 61″ | 42″ | 58″ | 39″ |
| Method of Tyring . . . | Hoop | Hoop | Hoop | Hoop | Hoop | Hoop |
| Width of Tyre . . . | 2½″ | 2½″ | 3″ | 3″ | 2½″ | 2½″ |
| Number of Spokes . . . | 14 | 12 | 14 | 12 | 14 | 12 |
| Track of Wheels . . . | 68″ | 68″ | 69″ | 69″ | 70″ | 70″ |
| **FRONT** | | | | | | |
| Maximum Width (Top) . . | 58″ | | 69″ | | 73″ | |
| Depth of Frontboard . . | 23″ | | 22″ | | 23″ | |
| Width of Sideboard . . | 9″ | | 8″ | | 7″ | |
| Width of Forebridge . . | 48″ | | 48″ | | 66″ | |
| **BODY** | | | | | | |
| Total Length (Top) . . | 127″ | | 132″ | | 112″ | |
| ,, ,, (Bottom) . . | 120″ | | 127″ | | 108″ | |
| Front (Top) to Ground . . | 70″ | | 65″ | | 66″ | |
| Mid Point (Top) to Ground) . | 70″ | | 65″ | | 66″ | |
| Rear (Top) to Ground . . | 70″ | | 65″ | | 66″ | |
| Front (Bottom) to Ground . | 44″ | | 42″ | | 41″ | |
| Mid Point (Bottom) to Ground . | 44″ | | 42″ | | 41″ | |
| Rear (Bottom) to Ground . | 44″ | | 42″ | | 41″ | |
| **COLOURS** | Brown Buff undercarriage | | Brown Buff undercarriage | | Blue. Buff undercarriage | |

## (2) EAST ANGLIAN WAGON

| Sible Hedingham, West Essex. Built c. 1850 | | Hadleigh, Suffolk. Built c. 1890 | | Reepham, Norfolk. Built c. 1870 | | Bishop's Stortford, Hertfordshire. Built c. 1865 | |
|---|---|---|---|---|---|---|---|
| REAR | FORE | REAR | FORE | REAR | FORE | REAR | FORE |
| 69″ | 49″ | 62″ | 43″ | 66″ | 45″ | 64″ | 47″ |
| Strakes | Strakes | Hoop | Hoop | Strakes | Strakes | Strakes | Strakes |
| 7 | 6 | — | — | 7 | 6 | 6 | 5 |
| 3½″ | 3½″ | 3″ | 3″ | 3″ | 3″ | 3½″ | 3½″ |
| 14 | 12 | 12 | 10 | 14 | 12 | 12 | 10 |
| 68″ | 65″ | 68″ | 66″ | 67″ | 67″ | 66″ | 64″ |
| 61″ | | 66″ | | 79″ | | 65″ | |
| 30″ | | 33″ | | 25″ | | 29″ | |
| 5½″ | | 8″ | | 9″ | | 6″ | |
| 54″ | | 52″ | | 50″ | | 45″ | |
| 159″ | | 152″ | | 148″ | | 145″ | |
| 147″ | | 144″ | | 135″ | | 138″ | |
| 79″ | | 84″ | | 68″ | | 74″ | |
| 69″ | | 70″ | | 65″ | | 68″ | |
| 76″ | | 84″ | | 73″ | | 74″ | |
| 42″ | | 45″ | | 41″ | | 42″ | |
| 40″ | | 43″ | | 42″ | | 41″ | |
| 42″ | | 45″ | | 46″ | | 43″ | |
| Blue Red undercarriage | | Blue Red undercarriage | | Stone Red undercarriage | | Blue Red undercarriage | |

## (2) RUTLAND WAGON

| St. Neots, Huntingdonshire. Built c. 1900 | | Exton, Rutland. Built c. 1860 | | Kingscliffe, Northamptonshire. Built c. 1835 | | Towcester, Northamptonshire. Built 1885 | | Near Rugby, Warwickshire. Built 1880 | |
|---|---|---|---|---|---|---|---|---|---|
| REAR | FORE | REAR | FORE | REAR | FORE | REAR | FORE | REAR | FORE |
| 57″ | 45″ | 59″ | 44″ | 57″ | 46″ | 56″ | 46″ | 59″ | 47″ |
| Hoop | Hoop | Hoop | Hoop | Hoop | Hoop | Hoop | Hoop | Hoop & Strakes | |
| 3″ | 3″ | 3″ | 3″ | 3″ | 3″ | 3″ | 3″ | 4½″ | 4½″ |
| 14 | 12 | 14 | 12 | 14 | 12 | 12 | 10 | 12 | 12 |
| 65″ | 65″ | 63″ | 63″ | 63″ | 63″ | 62″ | 62″ | 66″ | 66″ |
| 69″ | | 69″ | | 65″ | | 70″ | | 69″ | |
| 23″ | | 23″ | | 26″ | | 22″ | | 29″ | |
| 9″ | | 8″ | | 9″ | | 10″ | | 8″ | |
| 48″ | | 43″ | | 45″ | | 50″ | | 48″ | |
| 141″ | | 145″ | | 140″ | | 138″ | | 140″ | |
| 134″ | | 133″ | | 128″ | | 132″ | | 136″ | |
| 69″ | | 64″ | | 66″ | | 58″ | | 68″ | |
| 62″ | | 63″ | | 62″ | | 59″ | | 66″ | |
| 67″ | | 64″ | | 66″ | | 64″ | | 68″ | |
| 38″ | | 37″ | | 40″ | | 36″ | | 36″ | |
| 37″ | | 36″ | | 36″ | | 39″ | | 34″ | |
| 41″ | | 37″ | | 40″ | | 42″ | | 36″ | |
| Brick Red | | Brick Red | | Brick Red | | Brick Red Black lines | | Blue Red undercarriage | |

# BOX WAGONS—3 WEST MIDLANDS

## (a) SOUTHERN TYPE

| | Ledbury, Hereford. Built 1845 | | Leominster, Hereford. Built c. 1900 | | Powick, Worcestershire. Built c. 1880 | |
|---|---|---|---|---|---|---|
| WHEELS | REAR | FORE | REAR | FORE | REAR | FORE |
| Diameter . . . . | 59″ | 48″ | 56″ | 43″ | 51″ | 43″ |
| Method of Tyring . . . | Strakes | Strakes | Strakes | Strakes | Strakes | Strakes |
| Number of Strakes . . . | 7+7 | 5+5 | 6+6 | 5+5 | 6+6 | 5+5 |
| Width of Tyre . . . | 7″ | 7″ | 7″ | 7″ | 5″ | 5″ |
| Number of Spokes . . . | 12 | 10 | 12 | 10 | 12 | 10 |
| Track of Wheels . . . | 60″ | 60″ | 69″ | 69″ | 66″ | 66″ |
| **FRONT** | | | | | | |
| Maximum Width (Top) . . | 48″ | | 46″ | | 44″ | |
| Depth of Frontboard . . | 19″ | | 17½″ | | 20″ | |
| Width of Sideboard . . | — | | — | | — | |
| Width of Forebridge . . | 51″ | | 53″ | | 51″ | |
| **BODY** | | | | | | |
| Total Length (Top) . . | 156″ | | 151″ | | 131″ | |
| ,, ,, (Bottom) . . | 147″ | | 149″ | | 128″ | |
| Front (Top) to Ground . . | 64″ | | 65″ | | 68″ | |
| Mid Point (Top) to Ground . | 62″ | | 63″ | | 63″ | |
| Rear (Top) to Ground . . | 60″ | | 62″ | | 63″ | |
| Front (Bottom) to Ground . | 45″ | | 46″ | | 48″ | |
| Mid Point (Bottom) to Ground . | 43″ | | 43″ | | 43″ | |
| Rear (Bottom) to Ground . | 40″ | | 40″ | | 43″ | |
| COLOURS | Blue Red undercarriage | | Blue Red undercarriage | | Yellow Red undercarriage | |

## (b) CENTRAL TYPE

| | Kerry, Montgomeryshire. Built 1840s | | Onny Valley, Shropshire. Built 1915 | | Clun, Shropshire. Built c. 1880 | |
|---|---|---|---|---|---|---|
| WHEELS | REAR | FORE | REAR | FORE | REAR | FORE |
| Diameter . . . . | 58½″ | 46″ | 54″ | 47″ | 58″ | 44″ |
| Method of Tyring . . . | Strakes | Strakes | Strakes | Strakes | Strakes | Strakes |
| Number of Strakes . . . | 6+6 | 6+6 | 6+6 | 5+5 | 6+6 | 5+5 |
| Width of Tyre . . . | 6″ | 6″ | 5″ | 5″ | 6″ | 6″ |
| Number of Spokes . . . | 12 | 10 | 12 | 10 | 12 | 12 |
| Track of Wheels . . . | 60″ | 60″ | 62″ | 62″ | 62″ | 62″ |
| **FRONT** | | | | | | |
| Maximum Width (Top) . . | 66″ | | 69″ | | 69″ | |
| Depth of Frontboard . . | 21″ | | 22″ | | 20″ | |
| Width of Sideboard . . | 8″ | | 11″ | | 10″ | |
| Width of Forebridge . . | 46″ | | 50″ | | 48½″ | |
| **BODY** | | | | | | |
| Total Length (Top) . . | 146″ | | 151″ | | 156″ | |
| ,, ,, (Bottom) . . | 138″ | | 142″ | | 147″ | |
| Front (Top) to Ground . . | 68″ | | 72″ | | 69″ | |
| Mid Point (Top) to Ground . | 63″ | | 62″ | | 64″ | |
| Rear (Top) to Ground . . | 69½″ | | 66″ | | 70″ | |
| Front (Bottom) to Ground . | 42″ | | 45″ | | 45″ | |
| Mid Point (Bottom) to Ground . | 37″ | | 38″ | | 39″ | |
| Rear (Bottom) to Ground . | 48″ | | 42″ | | 46″ | |
| COLOURS | Yellow Red undercarriage | | Yellow Red undercarriage | | Yellow Red undercarriage | |

## (a) SOUTHERN TYPE

| Abergavenny, Monmouthshire. Built c. 1870 | | Newchurch Radnorshire. Built 1897 | |
|---|---|---|---|
| REAR | FORE | REAR | FORE |
| 56″ | 38″ | 57″ | 45″ |
| Strakes & Hoop | | Hoop | Hoop |
| 6 | 5 | — | — |
| 6½″ | 6½″ | 3″ | 3″ |
| 12 | 10 | 12 | 10 |
| 70″ | 70″ | 63″ | 63″ |
| | | | |
| 47″ | | 57″ | |
| 17″ | | 20″ | |
| — | | — | |
| 52″ | | 42″ | |
| | | | |
| 149″ | | 142″ | |
| 147″ | | 135″ | |
| 66½″ | | 75″ | |
| 58″ | | 64″ | |
| 60″ | | 68″ | |
| 42″ | | 45″ | |
| 38″ | | 38″ | |
| 37″ | | 45″ | |
| Blue Red undercarriage | | Prussian Blue Red undercarriage | |

## (b) CENTRAL TYPE

| Tregynnon, Montgomeryshire. Built c. 1890 | | Welshpool, Montgomeryshire. Built c. 1880 | | Llanbrynmair, Montgomeryshire. Built c. 1900 | |
|---|---|---|---|---|---|
| REAR | FORE | REAR | FORE | REAR | FORE |
| 53″ | 44″ | 58″ | 46″ | 57″ | 44″ |
| Strakes | Strakes | Strakes | Strakes | Strakes | Strakes |
| 6 | 5 | 6 | 5 | 6+6 | 5+5 |
| 3″ | 3″ | 4″ | 4″ | 6″ | 6″ |
| 12 | 10 | 12 | 12 | 12 | 10 |
| 62″ | 62″ | 63″ | 63″ | 63″ | 63″ |
| | | | | | |
| 60″ | | 63″ | | 64″ | |
| 24″ | | 22″ | | 26″ | |
| 6″ | | 12″ | | 8″ | |
| 44″ | | 47″ | | 43″ | |
| | | | | | |
| 139″ | | 148″ | | 147″ | |
| 128″ | | 138″ | | 138″ | |
| 64″ | | 80″ | | 62″ | |
| 53″ | | 63″ | | 54″ | |
| 57″ | | 70″ | | 63″ | |
| 40″ | | 47″ | | 40″ | |
| 33″ | | 41″ | | 34″ | |
| 41″ | | 49″ | | 45″ | |
| Yellow Red undercarriage | | Yellow Red undercarriage | | Yellow Red undercarriage | |

# BOX WAGONS—3 (*Continued*)

## (c) NORTHERN TYPE

| | Orton-on-the-Hill, Leicestershire. Built *c.* 1800-1820 | | Shifnal, Shropshire. Built *c.* 1840 | | Abbots Bromley, Staffordshire. Built *c.* 1845 | |
|---|---|---|---|---|---|---|
| **WHEELS** | REAR | FORE | REAR | FORE | REAR | FORE |
| Diameter . . . . | 55″ | 44″ | 57″ | 44″ | 58″ | 46″ |
| Method of Tyring . . . | Strakes | Strakes | Strakes | Strakes | Strakes | Strakes |
| Number of Strakes . . . | 6 | 5 | 6+6 | 5+5 | 6+6 | 5+5 |
| Width of Tyre . . . | 4″ | 4″ | 6″ | 6″ | 6″ | 6″ |
| Number of Spokes . . . | 12 | 10 | 12 | 10 | 12 | 10 |
| Track of Wheels . . . | App.60″ | App.60″ | 60″ | 60″ | 59″ | 59″ |
| **FRONT** | | | | | | |
| Maximum Width (Top) . . | 64″ | | 67″ | | 66″ | |
| Depth of Frontboard . . | 21″ | | 20″ | | 22″ | |
| Width of Sideboard . . | 13″ | | 12″ | | 11″ | |
| Width of Forebridge . . | 42″ | | 44″ | | 45″ | |
| **BODY** | | | | | | |
| Total Length (Top) . . | 147″ | | 156″ | | 150″ | |
| ,, ,, (Bottom) . . | 144″ | | 150″ | | 146″ | |
| Front (Top) to Ground . | 62″ | | 65″ | | — | |
| Rear (Top) to Ground . | 62″ | | 65″ | | — | |
| Mid Point (Top) to Ground . | 60″ | | 63″ | | — | |
| Front (Bottom) to Ground . | 40″ | | 44″ | | — | |
| Rear (Bottom) to Ground . | 40″ | | 44″ | | — | |
| Mid Point (Bottom) to Ground . | 33″ | | 37″ | | — | |
| **COLOURS** | Blue Red underparts | | Yellow Red underparts | | Blue Red underparts | |

# BOX WAGONS—4 SOUTH EASTERN

## (1) SUSSEX WAGON

| | Hove, Sussex Built 1880 | | Midhurst, Sussex Built *c.* 1880 | | Horsham, Sussex Built 1880 | |
|---|---|---|---|---|---|---|
| **WHEELS** | REAR | FORE | REAR | FORE | REAR | FORE |
| Diameter . . . . | 58″ | 37″ | 56″ | 37″ | 57″ | 48″ |
| Method of Tyring . . . | Hoop | Hoop | Strakes | Strakes | Hoop | Hoop |
| Number of Strakes . . . | — | — | 6+6 | 5+5 | — | — |
| Width of Tyre . . . | 4″ | 4″ | 6″ | 6″ | 3″ | 3″ |
| Number of Spokes . . . | 14 | 12 | 14 | 12 | 14 | 12 |
| Track of Wheels . . . | 62″ | 62″ | 70″ | 70″ | 65″ | 65″ |
| **FRONT** | | | | | | |
| Maximum Width (Top) . . | 63″ | | 68″ | | 63″ | |
| Depth of Frontboard . . | 18″ | | 20″ | | 22″ | |
| Width of Sideboard . . | 10″ | | 9″ | | 8″ | |
| Width of Forebridge . . | 40″ | | 43″ | | 40″ | |
| **BODY** | | | | | | |
| Total Length (Top) . . | 164″ | | 150″ | | 163″ | |
| ,, ,, (Bottom) . . | 144″ | | 130″ | | 144″ | |
| Front (Top) to Ground . | 65″ | | 67″ | | 68″ | |
| Mid Point (Top) to Ground . | 61″ | | 63″ | | 60″ | |
| Rear (Top) to Ground . | 71″ | | 68″ | | 67″ | |
| Front (Bottom) to Ground . | 37″ | | 40″ | | 39″ | |
| Mid Point (Bottom) to Ground . | 39″ | | 42″ | | 34″ | |
| Rear (Bottom) to Ground . | 35″ | | 40″ | | 39″ | |
| **COLOURS** | Blue Red undercarriage | | Blue Red undercarriage | | Blue Red undercarriage | |

## (c) NORTHERN TYPE

| | Wrexham, Denbighshire. Built c. 1880 | | Llangollen, Denbighshire. Built c. 1890 | |
|---|---|---|---|---|
| | REAR | FORE | REAR | FORE |
| | 60″ | 36″ | 57″ | 45″ |
| | Hoop | Hoop | Strakes | Strakes |
| | — | — | 6+6 | 5+5 |
| | 3″ | 3″ | 6″ | 6″ |
| | 14 | 12 | 12 | 12 |
| | 67″ | 67″ | 63″ | 63″ |
| | 65″ | | 67″ | |
| | 17½″ | | 24″ | |
| | 10″ | | 10″ | |
| | 45″ | | 43½″ | |
| | 142″ | | 144″ | |
| | 138″ | | 132″ | |
| | 55″ | | 59″ | |
| | 52″ | | 63″ | |
| | 53″ | | 74″ | |
| | 38″ | | 40″ | |
| | 36″ | | 44″ | |
| | 37″ | | 45″ | |
| | Red and Blue Red underparts | | Blue Red underparts | |

## (2) KENT WAGON

| | Dorking, Surrey Built c. 1850 | |
|---|---|---|
| | REAR | FORE |
| **WHEELS** | | |
| Diameter . . . . | 58″ | 46″ |
| Method of Tyring . . . | Strakes | Strakes |
| Width of Tyre . . . | 4″ | 4″ |
| Number of Spokes . . . | 14 | 12 |
| Track of Wheels . . . | 68″ | 68″ |
| **FRONT** | | |
| Maximum Width (Top) . . | 66″ | |
| Depth of Frontboard . . | 22″ | |
| Width of Sideboard . . | 9″ | |
| Width of Forebridge . . | 40″ | |
| **BODY** | | |
| Total Length (Top) . . | 157″ | |
| ,, ,, (Bottom) . . | 133″ | |
| Front (Top) to Ground . . | 69″ | |
| Mid Point (Top) to Ground . | 63″ | |
| Rear (Top) to Ground . . | 67″ | |
| Front (Bottom) to Ground . | 43″ | |
| Mid Point (Bottom) to Ground . | 44″ | |
| Rear (Bottom) to Ground . | 40″ | |
| **COLOURS** | Buff body and undercarriage | |

# BOX WAGONS—5 CENTRAL SOUTHERN ENGLAND
## (1) SURREY WAGON

| WHEELS | Alfold, Sussex. Built *c.* 1870 REAR | FORE | Selbourne, Hampshire. Built *c.* 1880 REAR | FORE | Rowlands Castle, Hampshire. Built *c.* 1860 REAR | FORE |
|---|---|---|---|---|---|---|
| Diameter . . . . | 56″ | 49″ | 59″ | 46″ | 56″ | 44″ |
| Method of Tyring . . . | Hoop | Hoop | Hoop | Hoop | Hoop | Hoop |
| Width of Tyre . . . | 3″ | 3″ | 2½″ | 2½″ | 2½″ | 2½″ |
| Number of Spokes . . . | 14 | 12 | 14 | 12 | 12 | 12 |
| Track of Wheels . . . | 56″ | 66″ | 74″ | 74″ | 77″ | 77″ |
| **FRONT** | | | | | | |
| Maximum Width (Top) . . | 72″ | | 72″ | | 68″ | |
| Depth of Frontboard . . | 18″ | | 19″ | | 16½″ | |
| Width of Sideboard . . | 9½″ | | 10″ | | 9½″ | |
| Width of Forebridge . . | 56″ | | 54″ | | 63″ | |
| **BODY** | | | | | | |
| Total Length (Top) . . | 147″ | | 153″ | | 135″ | |
| Total Length (Bottom) . . | 135″ | | 140″ | | 129″ | |
| Front (Top) to Ground . . | 65″ | | 67″ | | 61″ | |
| Mid Point (Top) to Ground . | 62″ | | 63″ | | 58″ | |
| Rear (Top) to Ground . . | 65″ | | 66″ | | 63″ | |
| Front (Bottom) to Ground . | 44″ | | 45″ | | 40″ | |
| Mid Point (Bottom) to Ground . | 40″ | | 41″ | | 36″ | |
| Rear (Bottom) to Ground . | 43″ | | 43″ | | 42″ | |
| **COLOURS** | Blue. Red undercarriage | | Buff. Light red undercarriage | | Brown. Red undercarriage | |

# BOX WAGONS—6 YORKSHIRE

| WHEELS | Wolds Wagon from Newbold. Built *c.* 1860 REAR | FORE | Dales Wagon from Bilsdale. Built *c.* 1895 REAR | FORE | Moor Wagon from Harkness Dales. Built *c.* 1850 REAR | FORE |
|---|---|---|---|---|---|---|
| Diameter . . . . | 60″ | 36″ | 50″ | 36″ | 54″ | 37″ |
| Method of Tyring . . . | Hoop | Hoop | Hoop | Hoop | Hoop | Hoop |
| Width of Tyre . . . | 3″ | 3″ | 3″ | 3″ | 3″ | 3″ |
| Number of Spokes . . . | 14 | 12 | 12 | 12 | 14 | 12 |
| Track of Wheels . . . | 67″ | 67″ | 50″ | 50″ | 58″ | 58″ |
| **FRONT** | | | | | | |
| Maximum Width (Top) . . | 65″ | | 60″ | | 60″ | |
| Depth of Frontboard . . | 17½″ | | 16″ | | 17″ | |
| Width of Sideboard . . | 10″ | | 9″ | | 9″ | |
| Width of Forebridge . . | 45″ | | 42″ | | 42″ | |
| **BODY** | | | | | | |
| Total Length (Top) . . | 142″ | | 100″ | | 120″ | |
| ,, ,, (Bottom) . . | 138″ | | 96″ | | 104″ | |
| Front (Top) to Ground . . | 55″ | | 46″ | | 54″ | |
| Mid Point (Top) to Ground . | 52″ | | 45″ | | 53″ | |
| Rear (Top) to Ground . . | 53″ | | 46″ | | 54″ | |
| Front (Bottom) to Ground . | 38″ | | 30″ | | 36″ | |
| Mid Point (Bottom) to Ground . | 36″ | | 30″ | | 35″ | |
| Rear (Bottom) to Ground . | 37″ | | 30″ | | 36″ | |
| **COLOURS** | Red and blue. Red underparts | | Brown. Red underparts | | Brown. Red underparts | |

## (2) DORSET WAGON

| Sherborne, Dorset. Built c. 1870 | | Sturminster Newton, Dorset. Built 1856 | | Plaitford, Hampshire. Built c. 1850 | |
|---|---|---|---|---|---|
| REAR | FORE | REAR | FORE | REAR | FORE |
| 53″ | 37″ | 46″ | 39″ | 55″ | 46″ |
| Hoop | Hoop | Hoop | Hoop | Hoop | Hoop |
| 2½″ | 2½″ | 2″ | 2″ | 2½″ | 2½″ |
| 12 | 10 | 12 | 10 | 12 | 10 |
| 71″ | 71″ | 71″ | 71″ | 68″ | 68″ |
| | | | | | |
| 70″ | | 70″ | | 73″ | |
| 16″ | | 17″ | | 18″ | |
| 9″ | | 8″ | | 9″ | |
| 57″ | | 57″ | | 56″ | |
| | | | | | |
| 134″ | | 118″ | | 135″ | |
| 124″ | | 111″ | | 125″ | |
| 60″ | | 56″ | | 61″ | |
| 54″ | | 52″ | | 58″ | |
| 58″ | | 57″ | | 65″ | |
| 38″ | | 37″ | | 39″ | |
| 38″ | | 37″ | | 38″ | |
| 38″ | | 37″ | | 39″ | |
| | | | | | |
| Yellow. Red undercarriage | | Yellow. Red undercarriage | | Blue. Red undercarriage | |

# BOW WAGONS—1 SOUTH MIDLANDS SPINDLE-SIDED

| | Eastleach Turville, Gloucestershire. Built 1880 | | Hailey, Oxfordshire. Built 1838 | | Thame, Oxfordshire. Built 1828 | |
|---|---|---|---|---|---|---|
| **WHEELS** | REAR | FORE | REAR | FORE | REAR | FORE |
| Diameter . . . . | 60″ | 51″ | 62″ | 51″ | 58″ | 48″ |
| Method of Tyring . . . | Hoop | Hoop | Strakes | Strakes | Treble Strakes | |
| Width of Tyre . . . | 2½″ | 2½″ | 3″ | 3″ | 8″ | 8″ |
| Number of Spokes . . . | 12 | 10 | 12 | 10 | 12 | 10 |
| Track of Wheels . . . | 64″ | 63″ | 62″ | 62″ | 66″ | 66″ |
| **FRONT** | | | | | | |
| Maximum Width (Top) . . | 76″ | | 69″ | | 78″ | |
| Depth of Frontboard . . | 15″ | | 14″ | | 18″ | |
| Width of Sideboard . . | 15½″ | | 17″ | | 14″ | |
| Width of Forebridge . . | 43″ | | 42″ | | 44″ | |
| **BODY** | | | | | | |
| Total Length (Top) . . | 164″ | | 156″ | | 161″ | |
| ,, ,, (Bottom) . . | 145″ | | 140″ | | 133″ | |
| Front (Top) to Ground . . | 67″ | | 69″ | | 69″ | |
| Bow (Top) to Ground . . | 68″ | | 64½″ | | 69″ | |
| Rear (Top) to Ground . . | 67″ | | 68″ | | 70″ | |
| Front (Bottom) to Ground . | 46″ | | 45″ | | 48″ | |
| Mid Point (Bottom) to Ground . | 44″ | | 40″ | | 38″ | |
| Rear (Bottom) to Ground . | 47″ | | 45″ | | 42″ | |
| **COLOURS** | Yellow. Red undercarriage, blck. & yell. linings | | Yellow. Red undercarriage, yellow linings | | Yellow. Red undercarriage | |

# BOW WAGONS—2 WESSEX AND LOWER
## (1) WILTSHIRE WAGON

| | Pewsham. Built c. 1840 | | Marlborough Built c. 1856 | | Near Salisbury Built c. 1870 | |
|---|---|---|---|---|---|---|
| **WHEELS** | REAR | FORE | REAR | FORE | REAR | FORE |
| Diameter . . . . | 56″ | 45″ | 58″ | 48″ | 59″ | 44″ |
| Method of Tyring . . . | Hoop & Strakes | | Strakes | Strakes | Hoop | Hoop |
| Width of Tyre . . . | 6″ | 6″ | 2½″ | 2½″ | 3″ | 3″ |
| Number of Spokes . . . | 12 | 12 | 12 | 12 | 12 | 10 |
| Track of Wheels . . . | 64½″ | 64½″ | 63″ | 63″ | 76″ | 76″ |
| **FRONT** | | | | | | |
| Maximum Width (Top) . . | 71″ | | 70″ | | 66½″ | |
| Depth of Frontboard . . | 15″ | | 14″ | | 15″ | |
| Width of Sideboard . . | 14″ | | 13″ | | 9″ (widening to 14″ at wheel bow) | |
| Width of Forebridge . . | 53″ | | 52″ | | 59″ | |
| **BODY** | | | | | | |
| Total Length (Top) . . | 152″ | | 156″ | | 150″ | |
| ,, ,, (Bottom) . . | 139″ | | 144″ | | 144″ | |
| Front (Top) to Ground . . | 64″ | | 65″ | | 60″ (approx.) | |
| Bow (Top) to Ground) . . | 63″ | | 66″ | | 63″ ,, | |
| Rear (Top) to Ground . . | 60″ | | 60″ | | 62″ ,, | |
| Front (Bottom) to Ground . | 48″ | | 51″ | | 45″ ,, | |
| Mid Point (Bottom) to Ground . | 39″ | | 50″ | | 47″ ,, | |
| Rear (Bottom) to Ground . | 41″ | | 51″ | | 47″ ,, | |
| **COLOURS** | Blue. Red undercarriage | | Blue. Red undercarriage | | Blue. Red undercarriage | |

| Whitchurch, Buckinghamshire. Built c. 1875 | | Shipston-on-Stour, Warwickshire. Built c. 1850 | |
|---|---|---|---|
| REAR | FORE | REAR | FORE |
| 58″ | 48″ | 60″ | 51″ |
| Hoop | Hoop | Strakes | Strakes |
| 2½″ | 2½″ | 2½″ | 2½″ |
| 12 | 10 | 12 | 10 |
| 64″ | 64″ | 72″ | 72″ |
| | | | |
| 71″ | | 71″ | |
| 19″ | | 22″ | |
| 13″ | | 12″ | |
| 42″ | | 44″ | |
| | | | |
| 156″ | | 168″ | |
| 140″ | | 148″ | |
| 69″ | | 75″ | |
| 69″ | | 69″ | |
| 64″ | | 69″ | |
| 40″ | | 45″ | |
| 39″ | | 42″ | |
| 40″ | | 43″ | |
| Yellow. Red undercarriage | | Yellow. Red undercarriage | |

## SEVERN BASIN PANEL-SIDED

| (2) W. BERKSHIRE. Lambourn, Berkshire. Built 1870s | | (3) N.W. HANTS. Monks Sherbourne, Hants. Built c. 1870 | | (4) DORSET BOW Romsey, Hants. Built c. 1870 | |
|---|---|---|---|---|---|
| REAR | FORE | REAR | FORE | REAR | FORE |
| 61″ | 50″ | 62″ | 48″ | 54″ | 41″ |
| Hoop | Hoop | Hoop | Hoop | Hoop | Hoop |
| 2½″ | 2½″ | 2½″ | 2½″ | 2½″ | 2½″ |
| 12 | 10 | 12 | 10 | 12 | 10 |
| 68″ | 68″ | 66″ | 66″ | 79″ | 79″ |
| 72″ | | 74″ (85″ across arches) | | 70″ | |
| 17½″ (14½″ in centre) | | 18″ | | 13″ | |
| 11″ | | 12″ | | 9″ | |
| 48″ | | 54″ | | 48″ | |
| 162″ | | 158″ | | 152″ | |
| 140″ | | 136″ | | 144″ | |
| 67″ | | 68″ (approx.) | | 62″ | |
| 67½″ | | 68″ ,, | | 61″ | |
| 68″ | | 68″ ,, | | 62″ | |
| 40″ | | 46″ ,, | | 44″ | |
| 42″ | | 44″ ,, | | 42″ | |
| 40″ | | 44″ ,, | | 42″ | |
| Yellow. Red undercarriage Red frontboard | | Blue. Red undercarriage | | Blue. Red undercarriage | |

## BOW WAGONS—2 (*Continued*)

### (5) N. SOMERSET & VALE OF BERKELEY

| WHEELS | Bedminster, Somerset. Built 1850 REAR | FORE | Oldbury Neate, Gloucestershire. Built c. 1870 REAR | FORE | Stinchcombe, Gloucestershire. Built c. 1865 REAR | FORE |
|---|---|---|---|---|---|---|
| Diameter . . . . | 56″ | 46″ | 55″ | 44″ | 54″ | 42″ |
| Method of Tyring . . . | Strakes | Strakes | Hoop & Strakes | | Hoop & Strakes | |
| Number of Strakes . . . | 6+6 | 6+6 | 6 | 6 | 6 | 6 |
| Width of Tyre . . . | 6″ | 6″ | 6″ | 6″ | 4½″ | 4½″ |
| Number of Spokes . . . | 12 | 12 | 12 | 10 | 12 | 10 |
| Track of Wheels . . . | 63″ | 63″ | 65″ | 65″ | 63″ | 63″ |
| **FRONT** | | | | | | |
| Maximum Width . . . | 75″ | | 66″ | | 68″ | |
| Depth of Frontboard . . | 13″ | | 17″ | | 13″ | |
| Width of Sideboard . . | 14″ | | 13″ | | 15″ | |
| Width of Forebridge . . | 44″ | | 40″ | | 48″ | |
| **BODY** | | | | | | |
| Total Length (Top) . . | 147″ | | 151″ | | 159″ | |
| ,, ,, (Bottom) . . | 135″ | | 140″ | | 147″ | |
| Front (Top) to Ground . . | 64″ | | 69″ | | 60″ | |
| Bow (Top) to Ground . . | 63″ | | 63″ | | 60″ | |
| Rear (Top) to Ground . . | 62″ | | 61″ | | 56″ | |
| Front (Bottom) to Ground . | 41″ | | 38″ | | 43″ | |
| Mid Point (Bottom) to Ground . | 39″ | | 35″ | | 33″ | |
| Rear (Bottom) to Ground . | 39″ | | 39″ | | 36″ | |
| **COLOURS** | Dark Blue. Red undercarriage. Red frontboard | | Dark Blue. Red undercarriage | | Dark Blue. Red undercarriage. Red frontboard | |

## BOW WAGONS—3 SOUTH-WESTERN

### (1) SOMERSET WAGON

| WHEELS | Muchelney. Built c. 1860 REAR | FORE | Easton-in-Gordano Built c. 1870 REAR | FORE | Ilchester Built c. 1880 REAR | FORE |
|---|---|---|---|---|---|---|
| Diameter . . . . | 58″ | 45″ | 54″ | 44″ | 50″ | 42″ |
| Method of Tyring . . . | Strakes | Strakes | Strakes | Strakes | Strakes | Strakes |
| Number of Strakes . . . | 6 | 5 | 6+6 | 5+5 | 6 | 5 |
| Width of Tyre . . . | 3″ | 3″ | 5½″ | 5½″ | 3″ | 3″ |
| Number of Spokes . . . | 12 | 10 | 12 | 12 | 12 | 12 |
| Track of Wheels . . . | 67″ | 66″ | 70″ | 70″ | 67″ | 65″ |
| **FRONT** | | | | | | |
| Maximum Width (Top) . . | 73″ | | 74″ | | 71″ | |
| Depth of Frontboard . . | 14″ | | 16″ | | 13″ | |
| Width of Sideboard . . | 10″ | | 11″ | | 9″ | |
| Width of Forebridge . . | 56″ | | 59″ | | 57″ | |
| **BODY** | | | | | | |
| Total Length (Top) . . | 141″ | | 143″ | | 133″ | |
| ,, ,, (Bottom) . . | 131″ | | 132″ | | 127″ | |
| Front (Top) to Ground . . | 57″ | | 57″ | | 61″ | |
| Bow (Top) to Ground . . | 65″ | | 67″ | | 68″ | |
| Rear (Top) to Ground . . | 66″ | | 67″ | | 66″ | |
| Front (Bottom) to Ground . | 43″ | | 44″ | | 40″ (approx.) | |
| Mid Point (Bottom) to Ground . | 36″ | | 40″ | | 38″ ,, | |
| Rear (Bottom) to Ground . | 40″ | | 44″ | | 40″ ,, | |
| **COLOURS** | Blue. Red underparts | | Blue. Red underparts | | Yellow. Red underparts | |

## (6) GLAMORGAN WAGON

| Llanfair Discoed, Monmouthshire. Built c. 1860 | | Cowbridge, Glamorgan. Built 1885 | | Llanishen, Glamorgan. Built 1895 | |
|---|---|---|---|---|---|
| REAR | FORE | REAR | FORE | REAR | FORE |
| 58" | 48" | 57" | 46" | 51" | 40" |
| Strakes | Strakes | Hoop | Hoop | Hoop | Hoop |
| 6+6 | 5+5 | — | — | — | — |
| 6" | 6" | 2¾" | 2¾" | 2½" | 2½" |
| 12 | 10 | 14 | 12 | 12 | 12 |
| 65" | 65" | 60" | 60" | 63" | 63" |
| 61" | | 67" | | 63" | |
| 15" | | 15½" | | 19" | |
| 10" | | 15" | | 15" | |
| 53" | | 44" | | 30" | |
| 162" | | 156" | | 148" | |
| 140" | | 126" | | 129" | |
| 72" | | 71" | | 65" | |
| 65" | | 65" | | 60" | |
| 64" | | 66" | | 60" | |
| 46" | | 42" | | 43" | |
| 35" | | 35" | | 34" | |
| 41" | | 46" | | 39" | |
| Blue. Red undercarriage | | Blue Red undercarriage | | Blue. Red undercarriage | |

## (2) DEVON WAGON    (3) CORNISH WAGON

| Oakhampton, Devon. Built 1870s | | Bow Wagon, Truro District. Built c. 1895 | | Trolley, Newquay District. Built c. 1900 | |
|---|---|---|---|---|---|
| REAR | FORE | REAR | FORE | REAR | FORE |
| 54" | 44" | 48" | 39" | 48" | 34" |
| Hoop | Hoop | Hoop | Hoop | Hoop | Hoop |
| — | — | — | — | — | — |
| 2½" | 2½" | 3" | 3" | 3" | 3" |
| 12 | 12 | 12 | 12 | 12 | 12 |
| 67" | 66" | 66" | 66" | 68" | 68" |
| 69" | | 56" | | 55" | |
| 16" | | 6" | | — | |
| 10" | | 13" | | 10" (over rear w.) | |
| 52" | | 45" | | 55" | |
| 132" | | 108" | | 120" | |
| 120" | | 102" | | 120" | |
| 59" | | 52" | | 51" | |
| 61" | | 57" | | 57" | |
| 61" | | 54" | | 51" | |
| 42" | | 35" | | — | |
| 24" | | 36" | | — | |
| 40" | | 36" | | — | |
| Blue. Red underparts | | Blue. Red underparts | | Blue. Red underparts | |

# APPENDIX II

## FEATURES OF CONSTRUCTION

### OF EACH

## WAGON TYPE

| REGIONAL TYPE OF WAGON | BODY | | | | | | *COLOURS | | |
|---|---|---|---|---|---|---|---|---|---|
| | Panelled | Spindled | Planked | No Mid-rail | One Midrail | Two Mid-rails | Yellow | Blue | Brown |
| Lincoln . . . | | X | | | X | | | X | |
| E. Anglian . . . | X | | | | X | | | X | |
| Hertford . . . | | X | | | | X | | | X |
| Rutland . . . | | X | | | | X | | | |
| Hereford Panel-Sided . | X | | | | | X | | X | |
| Hereford Plank-Sided . | | | X | X | | | | X | |
| Worcester . . . | | | X | X | | | X | | |
| Monmouth . . . | X | | | X | | | X | | |
| Radnor . . . | X | | | | X | | | X | |
| Shropshire . . . | X | | | | X | | X | | |
| Montgomery . . | X | | X | X | | | X | | |
| Stafford . . . | X | X | | | X | | X | X | |
| Denbigh . . . | X | | | | X | | X | X | |
| Sussex . . . . | X | | | | X | | | X | |
| Kent . . . . | X | | | | X | | | | |
| Surrey . . . . | | X | | | X | | | | X |
| Dorset Box . . . | | X | | | X | | X | X | |
| Yorkshire . . . | | | X | X | | | | | X |
| S. Midlands Spindle-Sided | | X | | X | | | X | | |
| Wiltshire . . . | X | | | X | | | | X | |
| W. Berkshire . . | X | | | X | | | X | | |
| N.W. Hampshire . | X | | | X | | | | X | |
| Dorset Bow . . . | X | | | | X | | | X | |
| N. Somerset & V. of Berkeley | X | | | | X | | | X | |
| Glamorgan . . . | X | | | | X | | | X | |
| Somerset . . . | X | X | | | X | | | X | |
| Devon . . . . | X | X | | | X | | | X | |
| Cornish . . . | | X | | X | | | | X | |

*Colours refer to the body onl·

| | | SIDEBOARDS | | | | SIDE CONSTRUCTION | | | COUPLING POLE | | | |
|---|---|---|---|---|---|---|---|---|---|---|---|---|
| Red | Stone | Removable | Solid | Spindled | Railed | Straight | Waisted Frame and Side Plank | Waisted Frame, Notched Side Plank | Straight | Curved | Pinned to Cross Bar | Unpinned |
| X | | X | | | | | | X | X | | | X |
| | X | X | | | | | | X | | X | X | |
| | | | | | | X | | | X | | | |
| X | | | | X | | X | | | X | | | X |
| | | X | | | | X | | | X | | | X |
| | | X | | | | X | | | X | | | X |
| | | | X | | | X | | | X | | | X |
| | | | X | | | X | | | X | | | X |
| | | X | | | | X | | | X | | | X |
| | | | X | | | X | | | X | | | X |
| | | | X | | | X | | | X | | | X |
| | | | X | | | | | | X | | | X |
| | | | X | | | | | | X | | | X |
| | | | X | | | | X | | | X | | X |
| | X | | X | | | | X | | | X | | X |
| | | | | | | X | | | | X | | X |
| | | | X | | | X | | | | X | | X |
| | | | X | | | X | | | | X | X | |
| | | | | X | | | X | | X | | | X |
| | | | X | | | X | | | | X | X | |
| | | | X | | | | X | | | X | X | |
| | | | | | X | | X | | | X | X | |
| | | | X | | | X | | | | X | | X |
| | | | X | | | X | | | | X | X | |
| | | | X | | | X | | | X | X | X | |
| | | | | X | | X | | | | X | X | |
| | | | | X | | X | | | | X | X | |
| | | | | X | | X | | | | X | X | |

wo colours for one type indicates alternatives.

# APPENDIX III

## WAGON TYPES IN EACH COUNTY

| | |
|---|---|
| BEDFORDSHIRE | Hertford type in south. Rutland type in the major part of the county. South Midlands Bow Wagon in western chalk fringe. |
| BERKSHIRE | Waisted Wiltshire type in west. South Midlands type in east. Surrey type in southern margins. |
| BUCKINGHAMSHIRE | South Midlands Wagon in most parts. Hertfordshire type in east. |
| CAMBRIDGESHIRE | East Anglian Wagon and hermaphrodite in east. Rutland Wagon in west. |
| CHESHIRE | Wagons not now found. Stafford type probably found in nineteenth century. |
| CORNWALL | A few small South-Western Bow Wagons and trolleys on larger farms. |
| CUMBERLAND | Wagons unknown. |
| DERBYSHIRE | A few Lincolnshire Wagons in low-lying parts and in dales. |
| DEVON | South-Western Bow Wagons in flatter parts. |
| DORSET | Dorset Box Wagons. Dorset Bow Wagons (mainly in north). A few South-Western Bow Wagons. |
| DURHAM | No wagons. |
| ISLE OF ELY | Wagons rare. East Anglian and hermaphrodite in east. Rutland Wagon in west. |
| ESSEX | East Anglian Wagon. Relatively rare in south. |
| GLOUCESTERSHIRE | South Midlands Spindle-Sided Bow Wagon in Cotswolds. Panel-sided Bow Wagon in Severn Valley. Hereford Wagon in north-west. |
| HAMPSHIRE | North-West Hampshire Wagon in north-west. Dorset Bow Wagon in south-west. Surrey Wagon in remainder of county. |

| | |
|---|---|
| HEREFORD | Hereford Panel-sided and Hereford Plank-sided Wagons throughout the county. Trolleys. |
| HERTFORDSHIRE | Hertfordshire Wagon. South Midlands Wagon in western fringe. |
| HUNTINGDONSHIRE | Rutland Wagon. |
| KENT | South-Eastern Box Wagon. |
| LANCASHIRE | No farm wagons. |
| LEICESTERSHIRE | Lincolnshire Wagon in Vale of Belvoir. Rutland Wagon in remainder of county. Wagons are rare in the west. Staffordshire Wagon found occasionally in north-west. |
| LINCOLNSHIRE | Lincolnshire Wagon and hermaphrodite throughout the county. |
| MIDDLESEX | No wagons now. Those that did occur in the past were of the Hertfordshire type. |
| NORFOLK | East Anglian Wagon. Smaller in size than those of Suffolk and Essex. Hermaphrodites. |
| NORTHAMPTONSHIRE | Rutland Wagon in most parts. South Midlands Wagon on southern fringe. |
| NORTHUMBERLAND | No wagons. |
| NOTTINGHAMSHIRE | Lincolnshire Wagon. Hermaphrodites. |
| OXFORDSHIRE | South Midlands Wagon in all parts. |
| RUTLAND | Rutland Wagon in all parts. Lincolnshire Wagon on larger farms in east. |
| SHROPSHIRE | Shropshire Wagon in all parts, but very rare in the north of the county. |
| SOMERSET | South-Western Bow Wagon in all parts. Wiltshire type north of Mendips. Dorset Box Wagon in south-east. |
| STAFFORDSHIRE | Staffordshire Wagon in non-industrial parts. |
| SUFFOLK | Large East Anglian Wagon. |
| SURREY | South-Eastern Box Wagon in cast. Surrey Wagon in west. |
| SUSSEX | South-Eastern Box Wagon. |

Q

| WARWICK | South Midland Bow Wagon on southern fringe. Rutland Wagon in south and east. Staffordshire Wagon in north and west. |
|---|---|
| WESTMORLAND | No wagons. |
| ISLE OF WIGHT | Wiltshire Wagon. |
| WILTSHIRE | Wiltshire Wagon. |
| WORCESTERSHIRE | Hereford Plank-sided Wagon. Trolleys. |
| YORKSHIRE | Yorkshire Wagon in east. Trolley in West Riding. Lincolnshire Wagon in southern periphery and possibly in Holderness. |
| ANGLESEY | Occasional wagon of the Denbighshire type. |
| BRECON | Radnorshire Wagon in more level parts. |
| CAERNARVON | No wagons in major part of county. Occasional Denbighshire Wagon on northern coast. |
| CARDIGAN | No wagons. |
| CARMARTHEN | Occasional degenerate Glamorgan type in southern part of the county. |
| DENBIGHSHIRE | Denbighshire Wagon in flatter parts. |
| FLINT | Denbighshire Wagon. |
| GLAMORGAN | Glamorgan Wagon in the Vale of Glamorgan. |
| MERIONETH | A few Montgomeryshire Wagons in southern fringes, near the Dovey valley. |
| MONMOUTH | Hereford Wagon in north. Glamorgan Wagon in south. |
| MONTGOMERY | Montgomeryshire Wagon in Severn and Dovey valleys. |
| PEMBROKE | Occasional degenerate Glamorgan type in south of county. |
| RADNOR | Radnorshire Wagon on all the larger farms. |

# APPENDIX IV

## CATALOGUE OF WAGONS IN MUSEUMS

### BRISTOL—Blaise Castle Folk Museum
Accession Number 7781
VALE OF BERKELEY WAGON
Purchased
A wagon built in 1840 and used at Gotherington, Gloucestershire. Colour: Blue.

### HALIFAX—Shibden Hall Folk Museum
Accession Number 1958/10
LINCOLNSHIRE WAGON
Donor: H. C. Halidane
A spindle-sided wagon built in the Wakefield district and used at Clarke Hall, Wakefield, Yorkshire. Colour: Blue.

Accession Number 1958/11
LINCOLNSHIRE WAGON
Donor: J. C. Halidane
A late plank-sided wagon used at Clarke Hall, Wakefield, Yorkshire. Colour: Red.

### HUDDERSFIELD—Tolson Memorial Museum
Accession Number A113.57
MILLER'S WAGON
Donor: G. Gill
A spindle-sided wagon of c. 1820 used at Staveley Mill, Knaresborough, Yorkshire. Colour: Red.

### LEICESTER—City Museum
RUTLAND WAGON
Purchased
A wagon of c. 1880 from Newton Linford, Leicestershire. Colour: Blue.

RUTLAND WAGON
Purchased
A wagon of 1854 used in the village of Manton, Rutland. Colour: Red.

### NORWICH—Bridewell Museum
Accession Number 288.956
EAST ANGLIAN WAGON
Donor: W. J. Eastwick
A wagon of c. 1830 from Little Melton, Norfolk. Colour: Blue.

Accession Number 8.957
EAST ANGLIAN WAGON
Purchased
A wagon of 1850 from Brooke, Norfolk. Colour: Stone.

PLYMOUTH—Buckland Abbey Museum
Accession Number B.A. 50.75
SOUTH MIDLANDS BOW WAGON
Donor: Viscount Astor
A wagon of *c.* 1860 from Sutton Courtenay, Berkshire. Colour: Yellow.

READING UNIVERSITY—Museum of English Rural Life
Accession Number 51/1286
WILTSHIRE WAGON
Donor: H. P. R. Print          December, 1951
This wagon was made in the village of Pewsham in Wiltshire by Holly, the village wheelwright. It was built *circa* 1840 and spent most of its working life on the Pullen family's farm at Potterne. The Pullen family later moved to Farmoor in Oxfordshire, taking this vehicle with them. Mr. Print bought the wagon at the sale in Farmoor before presenting it to the Museum.

Although the wagon bears the name-plate 'Humphries and Sons, Wagon Works, Chippenham', it is most unlikely that the wagon was made by them, since Edwin Humphries did not establish his wagon works at London Road, Chippenham, until 1857. The 'and Sons' could not possibly have been incorporated in the tradeplate before 1871, since before that date Humphries had only one son, Francis (born 1864). Another son, Albert Edward,[1] was born in 1871. All the evidence, including the style of the tradeplate itself, suggests that it was not fitted on the wagon until the end of the nineteenth century. The vehicle, which shows signs of extensive repairs, was probably taken to the Chippenham Wagon Works in the 1890s or early 1900s. Colour: Blue.

Accession Number 51/1295
MILLER'S WAGON
Purchased          December, 1951
This covered wagon was used for transporting grain and flour to and from the corn mill. It was made by Meadcroft, of Welwyn, for Benjamin Cole, of Codicote Mill, near Harpenden. The canvas tilt was made by Peddar, of Luton, before the First World War, the black patch on the side of the cover indicating the word 'Luton', which was blacked out as an invasion precaution. Colour: Yellow.

Accession Number 52/351
SOUTH MIDLANDS BOW WAGON
Purchased          November, 1952
This bow wagon was built at Eastleach Turville in the Cotswolds during the last quarter of the nineteenth century. It was made for Robert Howard who farmed in the locality. Colour: Yellow.

---

[1] I am indebted to Mr. K. P. Humphries of Cambridge for this information regarding to his family history.

## Accession Number 53/8
## CARRIER'S WAGON
Donor: T. W. Bagshawe     March, 1953

This heavy road wagon is believed to have been built in 1780, and it was used by a carrier called Webb, of Streetly End, in Cambridgeshire. The wagon itself was used to carry goods between Cambridge, Streetly End and London, the journey in each direction taking four days. At least four horses were required to draw this heavy vehicle, which was at one time fitted with a canvas cover. Colour: Blue.

## Accession Number 53/567
## LINCOLNSHIRE WAGON
Donor: Lt.-Col. W. Reeve     September, 1953

A spindle-sided box wagon built in 1829 for General John Reeve of the Grenadier Guards, who fought in the Peninsular War and at Waterloo. He was the great-grandfather of the donor and farmed at Leadenham in Lincolnshire. Colour: Prussian blue.

## Accession Number 53/655
## SOUTH MIDLANDS BOW WAGON
Donor: T. Harris     November, 1953

This wagon, which was built in 1838, was used at Hailey, near Witney in Oxfordshire. It was built for the donor's grandfather, Giles Harris, by the Charlbury wheelwright, Kench. The timber was found on Mr. Harris's farm, and the craftsman made it on the farm itself rather than in the village workshop. The wagon was used every year until 1951 and is in a perfect state of preservation. Colour: Yellow.

## Accession Number 54/365
## WEST BERKSHIRE WAGON
Purchased     May, 1954

This bow wagon, which is an improved version of the straight-framed Wiltshire wagon, was built for George Baylis, of Wyfield Manor, Boxford, near Newbury. Baylis was a well-known improver in the nineteenth century who, in the 1860s, started with a mixed farm of 350 acres and added holding after holding, until he was farming 12,000 acres for the production of wheat and barley alone. Colour: Yellow.

## Accession Number 54/677
## SUSSEX WAGON
Donor: Miss E. Godman     October, 1954

This wagon, an example of the narrow-wheeled variety of Sussex wagon, was made for the Godman family of South Lodge, Horsham, by a local craftsman. Colour: Blue.

## Accession Number 54/678
## SOUTH MIDLANDS WAGON
Donor: City of Gloucester Folk Museum.     September, 1954

This Cotswold wagon was presented to the Gloucester Folk Museum by Messrs. W. R. Haines and Sons, of Westington, Chipping Campden, and was probably built for them in the 1870s. Colour: Yellow.

Accession Number 55/74
## DORSET BOX WAGON
Donor: Miss J. Goodden          February, 1955

This spindle-sided box wagon was built *circa* 1870 near Sherborne in Dorset
and belonged to Lt.-Col. J. B. A. Goodden of Compton Hawy. It was probably
built by W. Hart, a wheelwright at Nether Compton, in Dorset, whose name
appears on the tailboard. The date when the wagon was last painted, 1937, is
also inscribed on the tailboard. Colour: Yellow.

Accession Number 55/501
## MONMOUTHSHIRE WAGON
Donor: W. R. Fowler          September, 1955

This broad-wheeled deep-bodied box wagon was probably made during the
last decade of the nineteenth century. It was built for the father of the donor,
who farmed at Tirley, Gloucestershire. The wagon was in constant use until
September, 1955. Colour: Yellow.

Accession Number 55/765
## MILLER'S WAGON
Donor: Miss M. M. Cherry          October, 1955

This canvas-covered miller's wagon was built in the 1860s for the donor's
grandfather, a miller at Bloxham Grove, Bodicote, near Banbury in Oxfordshire. The wagon was used by him and also by his son and grandson for
transporting corn and flour to and from the mill. Colour: Yellow.

Accession Number 56/188
## EAST ANGLIAN WAGON
Donor: J. T. Tricker          May, 1956

This very large box wagon, probably built in the last quarter of the nineteenth
century, was used by the donor on his farm at Hadleigh in Suffolk. The donor
bought it at a sale in the 1930s at Raydon, where it had been in use since it
was made. Colour: Blue.

Accession Number 56/242
## RUTLAND WAGON
Donor: C. F. Tebbutt          June, 1956

This box wagon was built about 1900 in the St. Neots district of Huntingdonshire. Colour: Red.

Accession Number 56/304
## DEVONSHIRE WAGON
Purchased          October, 1956

This small bow wagon was made at Upton, near Cullompton in Devon,
*circa* 1850. The Prince family, for whom it was made, later moved to East-
hampstead in Berkshire, and the wagon was used there for many years.
Colour: Blue.

Accession Number 55/795
## EAST ANGLIAN WAGON
Donor: J. W. Anstee          December, 1955

This wagon was made for C. E. Brewster at Maplestead in North Essex in
the 1850s. It was bought by a Mr. Beehag of Sible Hedingham, *circa* 1914.

When the wagon was found it was in a very bad state of preservation and it has been almost completely rebuilt in the Museum workshop. Colour: Blue.

### Accession Number 57/165
### VALE OF BERKELEY WAGON
Donors: Misses E. M. and D. M. Cobb        May, 1957

This wagon was made by Cleverdon and Son of Bedminster, near Bristol, around 1880. This firm of wheelwrights was in operation between 1830 and 1925 and many wagons of this type in northern Somerset and the Vale of Berkeley were built by them. This particular vehicle was built for the donors' father, who used it regularly on his farm in Dymock, in north-western Gloucestershire. The vehicle was last painted in 1932 when it was used to transport the body of the deceased Mr. Cobb to the cemetery. This use of wagons for carrying the body of the deceased owner during the funeral was a widespread custom in the West of England and South Wales. Colour: Blue.

### Accession Number 58/85
### SUSSEX WAGON
Donor: H. E. Philpot          November, 1958

This wagon was built in 1939 by W. J. Tedham, of Cripp's Corner, near Bodiam, Sussex, to the order of Mr. L. P. Haynes, who was then farm manager on Messrs. Guinness' hop farms. It was presented to Mr. Philpot in 1957 by Mr. H. R. Roberts, the present farm manager, and the vehicle was completely renovated by Frank Tedham, of Northiam, a cousin of the maker. Colour: Blue.

### Accession Number 59/100
### RUTLAND WAGON
Donor: H. Lawrence          February, 1959

This wagon, of the Rutland type, was built in the 1880s for the Morris family, of Yelvertoft, near Rugby. Mr. William Morris, J.P., gave up farming in 1957 and the wagon was sold to the donor at an auction sale. Colour: Red.

### Accession Number 59/101
### STAFFORDSHIRE WAGON
Donor: Major R. S. Dyott          February, 1959

This is an example of the now rare Staffordshire Wagon, but it differs slightly from the traditional type in that it is spindle, rather than panel, sided, while it also possesses a very slight waist. It was built around 1860, but its history is unknown. Colour: Yellow.

### Accession Number 59/219
### SHROPSHIRE WAGON
Donor: W. Medlicott          May, 1959

A wagon built around 1915 at Bishop's Castle and used until recently at Lydbury North. Colour: Yellow.

## ST. FAGANS—Welsh Folk Museum, National Museum of Wales
### Accession Number 57/331/1
### DENBIGHSHIRE WAGON
Donor: W. F. Francis

A panel-sided box wagon used at Plas y Ward, Runthun. Colour: Red.

Accession Number 57/331/2
### DENBIGHSHIRE WAGON
Donor: W. F. Francis
A panel-sided box wagon of late nineteenth century date used at Plas y Ward, Rhuthun. Colour: Red.

Accession Number 49/58
### RADNOR WAGON
Purchased
A wagon built in 1897 to attend Queen Victoria's Jubilee Celebrations. Used at Vuallt, Newchurch, Radnor. Colour: Prussian blue.

Accession Number 48/251
### GLAMORGAN WAGON
Purchased
A bow wagon of *circa* 1880 used at Tydraw, Pyle, Glamorgan. Colour: Blue.

Accession Number 37/12
### DENBIGHSHIRE WAGON
Purchased
A wagon of *circa* 1840 from Foxhall, Denbigh. Colour: Blue.

Accession Number 54/396
### GLAMORGAN WAGON
Donor: I. J. Llewellyn
A wagon built *circa* 1900 and used at Llanishen Fach Farm, Cardiff. Colour: Blue.

Accession Number 30/389
### GLAMORGAN WAGON
Purchased
A wagon built in the 1870s at Cowbridge and used at Broadway, St. Nicholas, Glamorgan. Colour: Blue.

Accession Number 56/451/1
### MONTGOMERYSHIRE WAGON
Purchased
A wagon of *circa* 1890 used at Plasauduon, Llanwnnog. Colour: Yellow.

## STRATFORD-UPON-AVON—Mary Arden's Agricultural Museum
Accession Number M.1953.4.4
### SOUTH MIDLANDS BOW WAGON
Purchased
A wagon of *circa* 1870 used at Wincote, Warwickshire. Colour: Yellow.

## YORK—Castle Museum
### YORKSHIRE WAGON
A wagon of *circa* 1880 from North Yorkshire. Colour: Brown.

## YORKSHIRE WAGON

A wagon of *circa* 1880 from Farndale, East Yorkshire. Colour: Brown.

## YORKSHIRE WAGON
### Donor: H. Hyde

A Wolds wagon of *circa* 1820 used at Snainton, North Yorkshire, and at Fraisthorpe, East Yorkshire. Colour: Brown.

# INDEX

Acts of Parliament 30, 32, 121
Adriatic 46
Adze 69, 70, *75*
Albania 46
Aleppo 46
America 36, 37, 57
Anatolian Plateau 3, 46
Andalusia 45
Anderson, Stanley *62*
Anglesey 49, 156, 234
Anvil 75, 79
Acquitaine 52
Aragon 45
Asia 3, 6, 22, 28, 37, 43-57
Auger 70, *76*
Austria 43
Avon, river 198
Axe 67, 70, 76
Axle 33, 34, 35, 41, 81-3, *82, 84, 88,* 111-2; Dorset 171; E. Anglia 121; Glamorgan 201; Hereford 135-6; Hertford 127; Lincoln 116; Rutland 131; Shropshire 147; Somerset 206; S. Midlands 185; Stafford 153; Surrey 166-7; Sussex 158-9; Wiltshire 190; Yorkshire 176
Axle arm 25, 28, 33-4, 80, 81, *82*, 83, 111

Balkans 55
Baltic States 36, 44
Bedfordshire 98, 125, 130, 232
Belgium 8, 45
Berg, Gösta 44
Berkeley, Vale of *96*, 103, 198ff, *200*, 235, 239
Berkshire 72, 73, *86*, *96*, 102, 165, 182, 183, 187, 194ff, *194*, 227, 232, 237
Best, Henry 174
Bevel 70, *75*
Bishop Stortford, Herts. 120
Blacksmith 66, 74, 100
Body 37-40, 56, 89-99; panel sided 12, 16, 38, 39, 40, 94, *95*, 123-4, 137-8, 141, 143, 145, 148-9, 151, 154, 156-7, 160-1, 188ff; plank sided 16, 37, 38-9, 40, 94, *95*, 109, 139-40, 174, 177-8; spindle sided 9, 16,
37, 38, 39, 40, 93-4, *95*, 118-9, 123-4, 128-9, 132-3, 154, 168, 172, 186-7
Berkshire 195; Denbigh 157; Devon 210-1; Dorset 172, 198; E. Anglia 123; Glamorgan 202-3; Hampshire 196-7; Hereford 138, 140; Hertford 128; Lincoln 118; Monmouth 143; Montgomery 151-2; Radnor 145-6; Rutland 132; Shropshire 148-9; Somerset 208-9; S. Midlands 186-7; Stafford 154-5; Surrey 168-9; Sussex 161; Vale of Berkeley 200; Wiltshire 192-3; Yorkshire 177-8
Body standards *214, 215*
Bohemia 5
Bolster 83, *84*, 85, 87, *214, 215*
Boring bar 80
Boulonnais 46
Bourbonnais 45
Bow saw *75*
Box 80-1; boxing 26; boxing engine *76*
Brace and bit 64, *76*
Bradawl 68
Braking 99-100
Berkshire 195; Dorset 173; E. Anglia 125; Glamorgan 204; Hereford 139; Hertford 129; Rutland 134; Shropshire 150; Somerset 209; S. Midlands 187; Stafford 155; Surrey 169; Sussex 162; Wiltshire 194; Yorkshire 178
Breconshire 144, 234
Bridle, *see Crooked stick*
Brie 46
Bronze Age 3, 4, 5, 57
Bruzz, *see Chisel, morticing*
Buckinghamshire 125, 183, 227, 232
Burwell, Lincs. 116

Caernarvonshire 2, 3, 4
Callipers 63, *76*
Callow End, Worcs. 142
Cambridgeshire 11, 37, 102, 105, 120, 130, 232
Cardiganshire 48, 104, 234
Carmarthenshire 234
Cart 6, 13, 27, 39, 43, 45, 47, 50, 51, 52, 53, 58, 99, 170, 208; box 7, 12, 15, 38, 55, 56, 98; ox 3, *4*, 13, 14,

34, 43, 46, 174, 211; scotch 16, 33, 38, 39, 53, 103, 109, 119, 125, 174, 179
Cave drawing 5
Central Southern England 49, 164, 217, 224-5
Chamfering 38, 105
Cheshire 17, 21, 49, 53, 152, 180, 232,
Childe, V. Gordon 23, 24
Chilterns 182
China 24
Chisel 63, 68, 75, 76, 81; boxing 80; morticing 66, 76, 80; turning 75
Church Stoke, Mont. 147
Cleveland, Yorks. 39
Clifton-on-Teme, Worcs. 142
Clout plate 82
Coach 27, 32, 35, 37
Collar, horse 6
Colours 104-11;
   Dorset 172-3; E. Anglia 124; Glamorgan 203; Hereford 138; Hertford 129; Kent 163; Lincoln 119; Monmouth 114; Radnor 146; Rutland 133; Shropshire 149; Somerset 209; S. Midlands 187; Stafford 155; Surrey 169; Sussex 162; Wiltshire 193; Yorkshire 178
Compasses 64, 65
Compass plane 70
Conestoga 36, 57
Constable, John 85-7
Construction 22ff; 59-113; regional types of 230-1
Conway, Caern. 156
Cornwall 6, 13, 14, 35, 47, 48, 51, 58, 210, 211ff, 212, 229
Cost of wagon 15, 20, 183
Cotswolds 11, 20, 39, 52, 93, 98, 101, 102, 104, 106, 110, 182, 198
Coupling pin 4
Coupling pole 4, 34, 84, 85, 87, 88, 214, 215
Cranborne Chase, Dorset 197
Crete 3, 24
Crimea 44
Crooked stick 67, 75
Cross bar 84, 90, 92
Cumberland 51, 232
Czechoslovakia 43
Czekanowski, Jan 3. 5

Dales, Yorkshire 174-5
Dalmatia 46
Dark Ages 26
Dartmoor 209
Davis, T. 212
Decoration 7, 104-11, 107, 116, 119, 124

Dejbjerg, Denmark 4, 5, 25, 33, 35
Denbighshire 14, 20, 96, 152, 155ff, 156, 223, 234, 239, 240
Denmark 4, 22, 23, 43, 55
Derbyshire 50, 114, 232
Devon 47, 48, 96, 100, 102, 105, 108, 169, 204, 208, 209ff, 210, 229, 232, 238
Dish of wheels 9, 12, 13, 27-32, 28, 34, 41, 64, 71, 75, 80, 90, 101, 111
Distribution 13, 43-58, 44, 49
Dimensions 217ff
Dog stick 99-100
Dorset 29, 41, 47, 52, 86, 96, 100, 101, 102, 105, 109, 110, 164, 165, 169ff, 170, 196, 197ff, 199, 208, 225, 232, 238
Dovey, river 48, 51-2
Dowelling 25
Draught pole, 3, 5, 6, 34, 35, 37
Drawknife 67, 70, 71, 75, 83, 89
Dray 36, 37
Drug bat, see Drag shoe
Drag shoe 100
Durham 50, 225, 232

East Anglia 7, 8, 11, 12, 37, 38, 39, 40, 52, 54, 81, 91, 93, 96, 99, 102, 104, 106, 107, 110, 114, 120
Eastern Counties 49, 114, 217, 218-9
East Midlands 11, 39, 40, 49, 93, 106, 108, 125ff, 217, 218-9
Edinburgh 39
Egypt 22, 24
Emilia, Italy 46
End board, see Tailboard
Epirus 46
Essex 39, 52, 53, 94, 119, 120, 219, 232, 238
Europe, 3-6, 24, 28, 33, 35, 36, 37, 38, 43-57, 44
Evans, E. Estyn 113
Evering, see Sideboard
Eye bolt 88

Faithorne, W. 9
Felloe 24, 25, 26, 27, 61, 68-71, 77, 79, 214, 215
Felloe horse 69
Felloe pattern 70, 75
Finland 44
Fitzherbert, Anthony 7
Flanders 43, 46
Flintshire 152, 234
Floorboards 14, 37, 91
Forebridge 84, 90, 92
Forecarriage 4, 6, 34, 35, 36, 38, 83-7, 86

Forest of Dean 142
Forez 45
Fox, Cyril 57, 213
Frame 89-91; triangular 3, 4, 5, 6, 34; Denbigh 156; Dorset 171-2, 198; E. Anglia 122; Glamorgan 202; Hampshire 196-7; Hereford 137; Hertford 128; Kent 163-4; Lincoln 117; Monmouth 143; Radnor 145; Rutland 132; S. Midlands 186; Surrey 167; Sussex 160; Shropshire 148; Somerset 207-8; Stafford 154; Wiltshire 191; Yorkshire 177
Frame saw, *see Bow saw*
France 8, 15, 37, 43, 45, 46, *47*, 55
Franche Comté 45, 53
Frontboard 94, 99, 105ff, *107*, *214*, *215*; decoration *107*, 116, 119, 124
Furnace *72*, 74-5

Gambo 6, 7, 53, 55, 146, 174
Gauge, spoke set 64-5, 67, 68, *75*; spoke tongue 68, *75*
Georgia 22
Germany 3, 24, 36, 43, 56, 57
Gevaudun 45
Gillingham, Norfolk *122*
Glamorgan 14, 48, 49, 54, 57, *96*, 188, 189, 200ff, *202*, *203*, 229, 234, 240
Glastonbury, Som. 25, 33
Gloucestershire 48, 101, 102, 103, 110, 142, 144, 182, 198, 226, 228, 232
Gouge 63, *76*, 80
Greece 22

Hames 6
Hampshire 19, 20-1, 54, 105 164, 165, 169, 195ff, *196*, 224, 225, 227, 232
Hand saw 70
Hama, Syria, 46
Harpenden, Herts. *126*
Harness 6, 9, 18, 34-7
Harvest 7, 10, 17, 19, 43, 45, 53, 91
Haudricourt, A. G. 18
Haute Auvergne 45
Hennel, Thomas 101
Herefordshire 12, 39, 40, 49, *69*, *86*, *96*, 103, 105, 109, 110, 111, 135ff, 136, 139ff, *139*, 220, 233
Hermaphrodite 120, *122*, 178-9
Hertfordshire 93, 105, 106, 120, 125, *126*, 130, 131, 218, 219, 233
Hertzegovina 46
Horncastle, Lincs. 115
Horse 5, 6, 10, 34, 36, 55, 89
Hound 35, *84*, 85, 87, *88*, *214*, *215*

Hound shutter *88*
Hub *see Nave*
Hudleston, N. A. 174
Hungary 4, 35, 43
Huntingdonshire 37, 130, 219, 233

Iceland 43
Igiel, Germany 6
India 43
Indonesia 43
Indus Valley 22
Ireland 23, 50
Iron Age 4, 24, 25, 26, 33, 35, 71
Isle of Ely 225, 232
Isle of Wight 234
Italy 22, 23, 46, 55

Jack plane 70, 71, *75*
Jarvis 67, *75*
Jeffries, Richard xi

Kamchatka 43
Kent *96*, 103, 105, 157, 163ff, *163* 165, 223
Kerry, Mont. 147
Kesteven, Lincs. 114
King pin *84*, 85, *88*, *92*

Lade board, *see Sideboard*
La Mancha 45
Ladders 97, 103-4
Lake District 50
Lancashire 50, 180, 233
Lathe 25, 63, *78*
Lausitz 3, 4
Langon, Sweden 4, *5*
Leicestershire 90, 93-4, 105, 114, 125, 130, 152, 153, 222, 233
Lettering 105-11
Limagne 45
Limousin 45
Linch pin 81, 82-3, *84*, *215*
Lincolnshire 12, 39, 50, *86*, 93, *96*, 98, 102, 103, 105, 108, 109, 111, 114ff, *115*, 131, 218, 233, 235, 237
Lindsey, Lincs. 114
Lisle, Edward 19
Llanbister, Radnor 144
Lock 5, 13, 38, 90-1, 101-2, 112
Loire, river 45
Lombardy 46
Lorraine 45
Low Countries 8, 9, 12, 19, 116
Luttrell Psalter 6, 27, 104

Mallet 68
Manufacturers 38-9
Marlborough Downs 194
Marshall, William 174, 179, 183, 212-3
Massif Central 46, 52, 56
Measuring wheel, see Traveller
Mediterranean 13, 22, 35, 36, 45, 55
Mendips 198, 205
Merionethshire 234
Mesopotamia 3, 22, 23, 33
Middle Ages 26, 27, 37
Middle East 3
Middlesex 125, 126, 233
Millevaches Plateau 45
Missenden, Bucks. 127
Model, clay 3
Mongolia 46
Monmouthshire 49, 135, 142ff, 144, 201, 229, 234, 238
Monnow, river 48
Montgomeryshire 51-2, 53, 150ff, 151, 220, 221, 234, 240,
Moors, Yorkshire 175
Morph, morphey, see Hermaphrodite
Morticing cradle 64
Morvan 45
Mowbray, Vale of 39
Museum, Bristol 235; Halifax 235; Huddersfield 235; Leicester 235; Norwich 235-6; Plymouth 236; Reading University xi, 90-110, 236-41; Stratford upon Avon 240; Welsh Folk 239; York Castle 240-1

Nave 9, 25, 36, 41, 61, 62-4, 66, 67, 80, 84, 111
Nave bands 63, 66, 81, 84, 105
Near East 3
Neb, see Draught pole
Neolithic Period 3
Netherlands 8, 9, 43, 45, 55, 56, 57
Newchurch, Radnor 144, 240
Newstead, Roxburgh. 25, 26
Norfolk 52, 53, 105, 119, 219, 233, 235, 236
Normandy 52, 56
Northamptonshire 90, 130, 183, 219, 233
Northumberland 50, 233
Norway 23, 44, 56
Nottinghamshire 103, 114, 120, 218 233

Odenburg, Hungary 4, 5, 35
Origins 3-58, 116
Orton-on-the-Hill, Leics. 153
Oxen 5, 6, 34, 35, 36, 45, 55, 158

Oxfordshire 39, 41, 74, 105, 182, 226, 233

Paint 105
Palaikastro, Crete 3
Paris Basin 56
Parkes, J. 18-9
Peate, Iorwerth C. 15, 19, 150-1
Pembrokeshire 48, 49, 234
Persia 24, 46
Pewsey, Vale of 191
Philippines 43
Picardy 45, 46, 53, 56
Piedmont 46
Pillow 84, 85, 88, 92, 214, 215
Pitt, William 153
Poland 5, 43, 54
Pole brace 88
Portugal 3, 15, 43
Pottery, see Odenburg
Purbeck, Isle of 205
Pyne, W. H. 10

Radnorshire 53, 54, 96, 135, 144ff, 145, 221, 234, 240
Rail, see Sideboard
Rail, front top 84; inner top 84, 93, 98, 214, 215; mid 39, 93, 94, 214; outer top 84, 98, 214, 215; top 12, 37, 56
Rath, rave, see Sideboard
Rear carriage 4, 34, 38, 83-7
Rhine, river 43, 56, 57
Rickman, John 29
Ring dog 69
Roads 30-2
Roller scotch 99
Roman Period 5. 6, 7, 25-6, 35, 37
Rouerge 45
Russia 43, 44, 46, 54
Rutland 90, 96, 104, 105, 114, 125, 130, 133, 219, 233, 235, 238, 239

St. Neots, Hunts. 37
Salisbury Plain 189, 196, 197
Salisbury, Wilts. 181
Samson 76, 79, 80
Sardinia 3
Savoy 45
Sawyer 69
Scandinavia 4, 5, 26, 35, 36, 43, 44, 55, 56
Scarfing down 75
Scotland 23, 51
Scribe 68
Severn, river 57, 108
Severn Basin, Lower 48, 49, 188ff, 198-9, 204, 217, 226-9

s'Hertogenbosch, Siege of 8, *9*
Shaft bar *84, 88*
Shafts 6, 34, 35, 36, 37, 85, 87-9
Shelvings, *see Sideboard*
Shoeing hole 76
Shropshire 11, 12, 39, 47, 48, 51, 53, 90, *96*, 105, 106, 138, 146ff, *150*, 220, 222, 233, 239
Sible Hedingham, Essex *120*
Sideboard 12, 13, 14, 19, 56, 94, 96-8, 103, *214, 215*
Side frame *84, 92, 214, 215*
Side planks 91-4, *214, 215*
Side support 14, *84, 92*, 94-6, *96, 97*, 98
Sinclair, John 49-50
Sind 4, 23
Sinkiang 46
Skid pan, *see Braking*
Skid pin, *see Drag shoe*
Sleaford, Lincs. 116
Sledge 51, 53-4, *73*
Sledge hammer 67, 75, 79
Slider bar 84, *88, 214, 215*
Small, James 29
Somerset 53, 72, *86, 96*, 108, 169, 197, 198ff, 204ff, *207*, 228, 233
South Africa 36, 37, 57
South Eastern England *49*, 157ff
South Midlands 48, *49, 96*, 110, 111, 182ff, 184, 194, 204, 217, 226-7, 326, 237, 240
South West England *49*, 204ff, 217, 228-9
Spain 3, 15, 33, 45
Splinter bar 85, *88, 214, 215,*
Spoke 9, 22-3, 24, 25, 26, 32, 63-4, 65, 66-8, 70-1, *214, 215*; mortice 25, 41, 64, 66, 67; tenon 25, 67; tongue 25
Spoke dog *62*, 70, *76*
Spokeshave 67, 70, 71, *75*, 77, 81, 83, 89
Squat, *see Braking*
Staffordshire xi, *86, 96*, 108, 152ff, 153, 222, 223, 239
Stock, *see Nave*
Strake 27, 61, 71, 72, 74, *84, 214, 215*
straking 78-80
Strake nail 79
Streetly End, Cambs. *11*
Strickland, H. E. 181
Strouter, *see Side support*
Structure 59, 215
Sturt, George 29, 83, 87-9, 91, 111, 165, 168
Summer *84*, 90, 91, *92*
Surboard, *see Sideboard*
Suffolk 52, 53, 102, 105, 119, 219, 233, 239

Surrey 89, *96*, 98, 103, 105, 126, 157, 164ff, 166, 194, 196, 223, 233
Sussex 11, 12, 83, 85, *86*, 91-3, *96*, 102, 103, 106, *107*, 111, 157, 158ff, *161*, 164, 165, 222, 224, 233, 237, 239
Sweden 4, 23, 44, 54
Switzerland 46
Syria 3, 46

Tabriz, Persia 46
Tailboard 98-9, 105ff, *214, 215*
Tebbutt, C. F. 109, 113
Tee square 64
Tenon saw 67
Tepe Gawra 3
Thames, river 194
Tilt 7
Touraine 45
Traces 36, 37
Track, width of 121
Traveller 74, *76*
Tring, Herts. 126
Trolley 17, 21, 141-2, *142*, 179-80, 211-2
Trottle car 50
Tuke, John 39
Tumbril, see *Cart, box*
Turkestan 46
Turkey 24
Tusser, Thomas 7, 19
Tyre 25, 27, 28, 31, 32, 61, 71-80; tyring 26, 29, 71-80, *72, 73*
Tyre bender 75
Tyre dog 75, *76*
Tyre tongs *72*, 75, *76*, 79
Tyring platform *72, 73*, 75, 78

Undercarriage 33, 35, 81-7, *88*, 105; Berkshire 195; Dorset 171, 198; E. Anglia 122; Glamorgan 201-2; Hampshire 196-7; Hereford 135-6; Hertford 127; Lincoln 116; Rutland 131; Shropshire 147-8; Somerset 206-7; S. Midlands 185-6; Stafford 153-4; Surrey 167; Sussex 159; Vale of Berkeley 199-200; Wiltshire 190-1; Yorkshire 176
Urals 37, 43
Usk, river 48

Van Hillegart *9*
Velay 45

Wagon, barge *16*, 21, 29; boat *17*, 21, 41, 83; bow 13, 14, 15, 16, *49, 84*, 182-213, *215*, 217, 218-9; box 12,

13, 15, 16, *49*, 114-81, *214*, 217, 218ff; carrier's 7, 9, *10*, *11*, 19, 27, 37, 41, 90, 98, 125-6, 237; ceremonial 3, 4, 5, 22, 24, 25, 33; miller's 235, 236, 238; regional types, *see also under county names* 10, 11; three-wheeled 46
Wain 6, 7, 13, *14*, 35, 48, 55, 174, 211
Waist 12, 90-1, *92*, *214*, *215*
Wales 6, 14, 23, 35, 48, 51, 54, 58, 100, *102*, 105, 108, 144-52, *145*, *151*, 155-7, *156*, *203*, 234
Warwickshire 108, 130, 131, 152, 182, 183, 219, 227, 234, 240
Weald 11, 157
Wearing plate 90, *215*
Weller, George 83, 111
Welshpool, Mont. 147
Wessex 48, 49, 188-204, 217, 226-9
West Midlands *49*, 134-57, *142*, 217, 220-3
Westmorland 234
Wheel 7, 9, 13, 16, 22, 25, 26, 27, 38, 41, 56, 101-2; broad 30, 31, 71; disc, *see solid*; narrow 9, 11, 30, 31; solid 3, 22, 27, 33;
Devon 211; Dorset 170-1; E. Anglia 121; Glamorgan 201; Hereford 135-6; Hertford 127; Kent 163; Lincoln 116; Montgomery 15; Rutland 131;
Shropshire 147; Somerset 205-6; S. Midlands 184-5; Stafford 153; Surrey 166; Sussex 158; Vale of Berkeley 199; Wiltshire 189-90; Yorkshire 176
Wheel arch *215*
Wheel pit 67
Wheel stool 70
Wheelwright 8, 25, 28, 29, 32, 38, 61, *62*, 74
   *see also Sturt, G. and Weller G.*
Wheelwrighting 10, 15-6, 24, 61-113
Whippletree 36
Wiltshire 17, 41, 54, *86*, 89, *96*, 98, 102, 104, 105, 106, 165, 172, 183, 189-94, *190*, 196, 197-8, 226, 234, 236
Wolds, Yorkshire 175-6
Woodstock Wagon 183-4
Worcestershire 17, 20, *96*, 110, 135, 140ff, *141*, 220, 234
Worgan, G. B. 211, 213
Wye, river 48

Yoke, 3, 5, 5,
Yorkshire 17, 35, 36, 39, *49*, 50, *86*, *96*, 105, 106, *107*, 114, 173ff, *175*, 217, 224, 234, 240, 241
Young, Arthur 39, 183